THE ROCK STAR
PRODUCT OWNER'S MANUAL

Real-World Advanced Agile Execution Strategies

MATTHEW KRAMER

This book is dedicated to my mother Oralee Kramer who sacrificed so much to give me the education that has allowed me the life that I lead today, and to Tanja Diamond who helped me turn this idea into a well written book that I'm proud to deliver into your hands.

TABLE OF CONTENTS

Foreword

Matt Kramer has produced a truly remarkable book because it doesn't sit up in the clouds talking about the theory of Agile Product Ownership; it gets down into the trenches and shares how to be successful in a company as a Product Owner dealing with the day to day challenges.

I coached, consulted, and trained people in traditional and agile delivery environments for two decades. My clients ranged from large-scale complex enterprise IT organizations to smaller boutique companies, which has given me ample experience dealing with Agile Product Ownership.

For the past two years, I have been exclusively a practitioner as a startup Co-Founder and Product Owner (radar4.ai). What I can now see very clearly, as a real-life practitioner, is that there is a dearth of practical advice for Product Owners that is not based on theory but truly based on experience.

Matt has honed his craft and spent significant time and effort crafting an effective book based on decades of experience in the trenches. If you are a Product Owner or work with them, you've come to the right place, a place to improve your skills in this essential role significantly.

Joseph Flahiff - CEO, Author
Being Agile in a Waterfall World:
A guide for complex organizations.

Introduction

As a discipline Agile does a great job training organizations and Scrum Masters in Agile theory, but without a great Product Owner the best trained teams will fail every time.

I've seen too many organizations go through three or four days of training in Agile, and then give a brand-new Product Owner three or four sentences about how to do their brand-new job. At F5 Networks I work with the critical core engine teams for a multi-billion dollar a year product.

I've seen Agile programs flail about because no one spent the time to properly train the Product Owners, who are critical to the success of everyone. So, I decided to be part of the solution and share with the path to success with you.

HOW TO READ THIS BOOK

This book exists for two reasons:

Reason 1: This book is here to get you great results by giving you the information you really need to be successful on the ground, at the actual job.

Reason 2: It's here to make the standard Scrum certification course look as sad and inadequate as it really is, so that when you need training, you'll think of me and the follow-on course that has been created from this book (available on-line or in person).

In this book I'm going to be breaking a lot of the unwritten rules by being very direct and helping you understand how to succeed in the real world as a Product Owner. This is the information that the large consulting companies will only trickle out to you through onsite coaches who will drag the engagements out as long as possible. I don't have a legion of slick sales people, and I don't ever want to. I'm here to help you succeed.

Here is the situation that we're trying to avoid.

As a new Product Owner, you're leading a high-profile team responsible for delivering new content for an important client who brings in ten-million a month in revenue. Your team has been working hard and there are two hundred plus people on the zoom including your boss and the CEO so it's your time to shine.

You start showing off the brilliant new functionality that the team has been working on for the past three sprints when halfway through the demonstration the application crashes, and you notice a glaring spelling error in the product. When you finish, the director of sales mentions that he didn't see critical functionality promised to the customer. Your victory lap has been crushed in front of two hundred people, your boss is angry, and the Development Team that worked late nights to get this across the finish line is now demoralized.

I've seen this and worse at companies that I've worked for. Fortunately, this situation is avoidable and it's why I've written this book. This book is a practical guide for current and future Product Owners on succeeding in the real-world, day-to-day execution of the PO role. In short, this book will teach you how to be a Rock Star Product Owner.

What this book is not

There are plenty of books and training material about the basic Agile theory and concepts. This book does not attempt to recreate that information.

Instead, we'll go beyond basic Agile theory to look at successful Product Owner strategies with real-world examples.

What this book is

An Agile team might have the world's best Scrum Master, but without a good Product Owner, it's never going to deliver what the customer/company needs, and the customers will get gobbled up by the competition.

Agile training programs focus on the Scrum Master, or Agile itself, and people thrust into the PO role don't get the education required to succeed. Don't hope for success, read this book and go and get it.

I've spent decades training and working alongside Product Owners, both mediocre and exceptional. This book is the culmination of that experience. I will explain why the Product Owner role was created and how to succeed in your role as one. I'll also show you what development teams, executives, sales, and the business side of an organization need from you, the Product Owner, to be successful.

This book provides practical details on how Product Owners approach essential day-to-day tasks of the job. It teaches how to deliver amazing Sprint Demonstrations that inspire executive confidence.

You will learn to deal with impossible requests, the ones you can't realistically deliver on time, and yet you've been tasked with doing so.

I will provide detailed information about goals, strategies, and pitfalls to avoid for every Agile ceremony and primary interaction Product Owners will face.

Let's get started!

Product Owners

To deliver on the promise, about goals, strategies, and pitfalls to avoid for every Agile ceremony and primary interaction Product Owners will face, I can't just tell you what Rock Star Product Owners do; I also need to help you understand why they do what they do and how the role came into being. I have to walk the line between theory and practice. but trust me, we spend most of our time on the practical practices of Product Owners. A proper background will help you go beyond what you need to do in any given situation and customize your approach to get the best results.

What is a Product Owner?

Within a product, the Product Owner (PO) is responsible for prioritizing and overseeing the overall quality and creating features and functionality that will be valuable for internal or external customers.

The PO must balance business and team needs, make critical decisions and tradeoffs, decide what to build, how well to build it, and when it's complete enough to focus the team's attention on what's next.

Without a Product Owner

Without a Product Owner (or without a good Product Owner), the quality needed by the customer (internal or the end-customer) rarely gets defined with enough detail, so teams will deliver poor quality work that needs to be fixed and revised.

Without a Product Owner, Development Teams don't get the details to understand what needs to be built. Without enough details agile teams will build the wrong/incomplete features into the product which will require time and effort for them to rip it out or correct it once they realize the mistake.

Without a Product Owner, there isn't a knowledgeable person to review and accept or reject the work based on the customer's needs, so often what was asked for doesn't get delivered.

Without a Product Owner, The Business doesn't have anyone they can easily coordinate with when changes or new features are needed, so high priority customers have to wait for their features, and sometimes get new functionality that doesn't match what they asked for.

PART 1- CORE AGILE CEREMONIES & CORE CRITICAL DELIVERY

Agile Ceremonies

Agile Ceremonies are formal meetings held during critical times in the agile lifecycle, with set deliverables designed to ensure a positive and productive outcome.

The Product Owner has a central and essential role during the ceremonies that make up the Agile lifecycle. These ceremonies are critical to the entire team's success, and executing well in these events will ensure the project's successful completion.

The Agile ceremonies are significant milestones in the Agile process and are just as important to the successful progress and eventual completion of a project as the daily coding and testing is. The following is a detailed guide that explains how successful Product Owners approach each ceremony, the goals for each event, what to avoid, and how to execute.

The Ceremony/Situation

The Ceremony/Situation: Daily Standup, also known as the Daily Scrum or Scrum.

Agile description: A short daily planning session with the Scrum Master, Product Owner, and the Scrum Team brings visibility to the team's progress and ensures that anyone encountering issues or a lack of information can get help. Participants usually touch on what they did yesterday, what they plan to do today, and if they are running into any issues or need information.

The PO's primary role: Participant, be a resource for the team, and observer for the business.

The PO's secondary role: to act as backup for the Scrum Master.

Your goals

1. Ensure your team can be as productive as possible for the day by ensuring that:

   a. There are enough user stories and work for the team to make meaningful progress

   b. No one is stuck or needs help

   c. Acceptance criteria for Stories and Epics are clear

    d.   Answering questions that the team might have about what they should prioritize and focus on

2. Accept or reject any stories/work completed by the team, and provide guidance on why stories are rejected

3. Identify and mitigate newly emerging risks

4. Understand any impacts to the team's work

How to execute

1. Listen carefully to the team

2. Give the team space to carry on most of the conversation

3. When necessary, ask questions to understand the progress of the work being undertaken

4. Listen for engineers who might be stuck and need help

5. Give specifics around how you will attempt to get engineers the assistance or information when they need it

6. Hold any conversations that might take too long after the Standup. "Let's parking-lot this topic and pick it up after Standup."

Video/Zoom considerations

1. Use and encourage your team to use video to deepen relationships and open communication

2. Look for body language that is tentative or doesn't match what's being said

3. When possible socialize before and after the meeting to build team cohesion

What to avoid

1. Don't keep engineers to the short format of "What I did yesterday, what I'll do today, and help needed." This format is too short and doesn't provide the needed visibility. You and everyone else on the team need to get the details to understand what is happening

2. Don't dominate the conversation; this is the team's ceremony

3. If possible, don't skip this ceremony as you will learn what is going on and what help might be needed

4. Don't assign work to engineers. How work gets done is owned by the team. Even if it's difficult, they need to figure out how to do it

The Ceremony/Situation: Backlog Grooming

The PO's primary role: present and facilitate discussion of feature, user story drafts, and capture details to get them to their final form.

The PO's secondary role: NA

Agile description: These are working sessions to improve and socialize User Stories which are likely to be worked on in the near future. Use these to refresh old, out-of-date User Stories that were groomed too long ago (if older than a Sprint and a half, I recommend refreshing them).

Your goals

1. Continue to focus the team on the overarching purpose of the product

2. Introduce the Epics and User Stories that make up the Epics with everyone on the team

3. Identify any additional User Stories needed

4. Improve the User Stories by:

    a. Ensuring the team understands the User Story or Epics goal based on the title

    b. Editing the Acceptance Criteria for clarity based on team feedback

    c.   Capturing any assumptions or edits the team suggests

5.   Have enough Epics and User Stories ready for the next Sprint

How to execute

1.   Hold these meetings weekly to ensure that there are enough user stories and to keep them from becoming too long

2.   Send out the Epics and Stories beforehand, so people have time to review them

3.   Add the necessary details to the Epics and User Stories before sending them for review

4.   Be open to the team's suggestions and edits

Video/Zoom considerations

1.   Use video to allow for better communication

2.   Ask if the team understands each epic or story and watch for body language saying that someone might not

What to avoid

1. Don't cancel Grooming because the team's need for User Stories can pile up quickly.

2. Grooming stories that aren't ready

What can go wrong

1. You may discover that key features aren't defined well enough, or the design proposed by architects is flawed and won't work. Don't try to hide or sugar-coat these challenges. Take the following steps (in order):

    a. Identify the people and information you need to clear up this challenge

    b. Set up working meetings to go over and do the work

    c. Announce to your leadership the challenge and the plan you have to solve it

    d. Acknowledge the challenges the team will have without the correct information.

The Ceremony/Situation: Sprint Review, also known as a Sprint Demo

The PO's primary role: Owner and Master of Ceremonies.

The PO's secondary role: NA

Agile description: Agile teams demonstrate to stakeholders what they have completed during the Sprint.

Your goals

1. Instill confidence in your skills as a Product Owner:

    a. Demonstrate how your team delivered on its goals

    b. Show the value your team was able to create

    c. Show off working code and functionality as close to a production-like environment as possible

    d. Detail any risks that might endanger your current or future commitments and what mitigations are available

    e. Talk about the goals for the next Sprint

2. Instill confidence in your engineers and their ability to work together

    a. Highlight the hard work and technical competence of your team

3. Instill confidence in the stakeholders that you are a good Product Owner

    a. Highlight the value you created

    b. Demonstrate an understanding of the current product priorities

    c. Show your understanding about what the team should work on next

4. Enjoy the competence and achievements that you have helped define and create

5. Boost the moral, confidence, and motivation of the team by giving them main stage recognition for all of the hard work that they have completed (this is critical for your future success)

How to execute

1. Create an attractive Sprint Review/Demo template and use it every time (this is detailed and another section of the book)

2. Ask the team prior to the demo what risks they foresee, and share these during your presentation. This not only helps identify risks that leadership may be unaware of, but it also clearly communicates to the team that you hear their concerns and respect their opinions

3. Get your team to think about what they want to demo several days beforehand

4. Have the engineers prepare and conduct a dry-run of the demo the day before the presentation (still is critical to being able to give a smooth and polished presentation)

5. Include ways to mitigate risks in your presentation (if risks can be mitigated)

6. Create an email template to send out after the Sprint Demonstration highlighting all of the content and team accomplishments

7. Send an email recap to your boss, executives in your program, the sprint teams, and any stakeholders after the demo

Video/Zoom considerations

1. Wear professional clothes

2. Talk slower and put emphasis on the important points

3. Practice beforehand

4. Have notes to work from so you can give a good presentation via video

What to avoid

1. Don't try to show work or code that isn't easily demonstrable. Instead, offer a few screenshots of code and talk through what the team accomplished; tell them how it will improve the end-customers experience

2. Don't try to take the credit for your team's accomplishments. Giving credit where it's due helps motivate your team and shows Product Owner leadership. Deflect complements and praise from yourself back to the team

3. As much as you deflect praise and give it to the team, you absorb criticism. If there is a criticism of the product shown in the demo, own it

4. Don't skip talking about risks. Visibility into what could go wrong is essential in the role of a successful Product Owner

5. Don't miss showcasing what your team will be working on in the next Sprint. A good presentation will highlight the importance and value of what your team will deliver at the next demo

What can go wrong

1. If the code or functionality crashes during the presentation, keep calm and continue speaking while the issue is resolved. If the issue can't be resolved, explain the work, value, and functionality that is complete

2. Audio and presentation problems are common issues encountered during demonstrations. Conduct a quick practice with a one or two-person audience to ensure that your presentation can be seen and heard

The Ceremony/Situation: Sprint Planning, also known as Sprint Kickoff

Agile description: A ceremony to review and agree upon the Sprint Goals, the priorities for the team, and the User Stories to be worked on during the next Sprint.

Your goals

1. Get everyone on the team on board with the Goal for the Sprint and what the priorities are

2. Review any new User Stories to be included in the Sprint

3. Go over any new edits or changes to User Stories or Epics

How to execute

1. Send out a prioritized list of the User Stories before the meeting

2. Complete any updates identified during User Story Grooming

3. Know what is essential to and should be the Sprint Goal for the team

4. Tell the team what the most critical thing to be accomplished is and make this the Sprint Goal

Video/Zoom considerations

1.   Encourage the team to meet using video

2.   Carefully present the sprint goals at the beginning of our presentation, go into detail, and then recap with the sprint goals

3.   Having a written slide with sprint goals and a list of the epics/features would be useful

What to avoid

1.   Not having needed edits done before the meeting

2.   Not knowing what the top priority is

What can go wrong

1.   If there are stories that aren't complete and ready to be worked on, focus on getting enough work to engage everyone on the team and finish prepping the stories as soon as possible. Don't start work on incomplete stories

2.   New information may come up which invalidates User Stories or features your team was about to work on. If this happens, acknowledge the new situation to the team and take the steps identified above

The Ceremony/Situation: Sprint Retrospective

Agile description: The team gets together to understand what is going well and should be preserved, areas that could be improved, and which actions the team will take to improve one or two of the most important ones. Product Owners often don't have an invite to this ceremony, but it's helpful to attend if they can do so without disturbing the team's process of self-improvement.

Your goals

1. Listen and do not disturb the team's self-improvement process

2. Gain a deeper understanding of the team's strengths and weaknesses

3. Identify areas where you might adjust to help the team

How to execute

1. After the conversation has slowed, bring up any large items the team may have missed

2. Accept the team's decision about what they want to focus on

Video/Zoom considerations

1. If this meeting is being done via Zoom the team will need to have a way to list ideas and vote on them. Trello or other online tools that allow for dot voting are good options

What to avoid

1. Dominating the conversation

2. Trying to direct the team about what they should focus on

3. Telling the team how to fix something

What can go wrong

1. Team members may get into conflict. The Scrum Master should help focus this energy away from personal attacks and finger-pointing and towards identifying a productive solution. If the Scrum Master can't keep emotions in check, the Product Owner can help focus the conversation on solutions

2. If the issues brought up are items the team doesn't have direct control over, the Product Owner should examine problems and take ownership of them when appropriate

PART 2 - ESSENTIAL SKILLS

Understand your customer

Most products have a number of customers: Users, Administrators, internal customers, reporting, maintenance, to name a few. You will need to focus on who the customer is, how they will be using the product, and what they will find valuable. internal customers who need reporting and customer support features, business analytics, and other business-critical capabilities. Having a close relationship with The Business is essential to understand their strategic initiatives and priorities for delivering what they need to market, maintain, and monetize your product.

Your product will likely have multiple types of customers using it. Each customer type will have other own goals, strengths, and weaknesses. Don't combine the customer types into one because you will miss many details.

For example, as I'm writing this book, I need to keep in mind that different audiences with different backgrounds will read this book. There will be people who have never been a Product

Owner before, people who are currently Product Owners but wish to improve their expertise, and people who manage groups of Product Owners.

A typical and effective strategy for tracking different customer types is creating and naming a customer persona for each. For each customer persona, record their background, goals, how they will want to use your product and anything else that might influence their approach to using your product. It is often useful to give the person a name in addition to other demographic and psychographic data. You will be surprised at the insights you gain by building a feature for Judy, who is a thirty-two-year-old mom, rather than building it for PERSONA PROFILE TWO.

Having these personas in common usage by your teams will provide a clear vision of each customer type for consideration when building your product. Once personas are used long enough, your teams will remember them by name, keeping them in mind when making decisions or trade-offs.

Value

The Product Owners main goal is to create customer value. Within organizations it is often challenging to identify how valuable something is because there aren't process to quantify and compare value.

Further, there often isn't good communication with customers making it difficult to get feedback.

These problems are caused by:

1. Sales and Marketing departments which have a lot of input into deciding priorities, but have no incentives to engage to understand customers priorities (they are too busy marketing and selling.

2. No defined ownership for those who should be engaging with customer to get data about customer value.

3. No customer Profiles, leading to an incomplete understanding of who is using the product.

4. Feedback being gathered from the customer who complain the most, who may not be represent the majority opinion of your customers.

5. Customer "Experts" with large egos who are sure they know it all and are unwilling to ask a statistically relevant sampling of customers, sometimes not engaging customers at all.

6. No process or tool to score, rank or quantify proposed features.

7. A disconnect between people in the company getting the customer data, when it's being collected and the people making priority decisions about what should be built.

8. The prioritization of features is being done based on political or organizational power and not based on what the customer needs.

The good news is that there are appropriate tools and processes. I see tools and their processes can help establish the true customer value of proposed features. If these items don't exist then it would be useful for the Product Owner to introduce them and the benefits they would bring to the organization. Here are the items and how they are helpful.

A Ranked Feature/Attribute Comparison – A scorecard used to compare the major attributes and features of your product and how it compares to your competition.

Use: to understand how your product compares to others and which features might be useful to ask customers about. A one through three scale using red, yellow, and green can be a good way to visualize the quality and completeness of a feature.

Feature Scorecard – A scorecard which quantifies the value to the company for a new feature vs their level of effort. This is a powerful tool and especially useful is there are different groups competition to get new features prioritized. The scoring can be adjusted based on the business environment.

Use: Scores are assigned to propose features based on their positive impact or negative impact, and the levels of effort it will take to implement them.

- Legal Liability – 50 points
- Regulatory or compliance – 50 points
- Will cause a system outrage – 50 points
- Will block primary functionality – 30 points

- Will drive new revenue – 18 points
- Feature to match or move beyond competitor – 10 points
- Improvement in usability or clarity – 5 points
- Fix to the stabilization of the system – 5 points

Customer Surveys – Questionnaires used to ask customer about their use of your product.

Use: These surveys provide direct feedback from customer so that your organization can get beyond guessing and opinions to really understand what your customers want and how they feel about your product.

Here are some of the major survey types and their uses:

- Net Promoter Score (NPS) survey- How likely is a customer to recommend you to someone they know?

- Customer Satisfaction Score (CSAT) Survey – How happy or unhappy is a customer with your product?

- Customer Effort Score (CES) survey – a score for how much effort a customer has to exert to get an issue resolved, a request fulfilled, or a question answered.

- User Interface Effort Score (UIES) survey – a score for how much effort a customer has to exert in order to accomplish their goals using your product.

A Product Roadmap

A document supported and updated regularly which provides a single source to describe the features and improvement that will be undertaken for a product.

Use: Having a roadmap encourage feature decisions and prioritization to happen in a more holistic approach because everything is listed out in one location. Having a regular review meeting encourages stakeholder to define and elaborate the value of proposed feature so that the resources get allocated to features they care about.

How to Use Value –

Once the value of a given feature is understood the PO (or in some cases the organization) can make intelligent decisions about how to maximize the value of what will be delivered. By balancing easy to deliver items that have medium to high value, with longer horizon items that are must have features, a regular steady flow of customer value can be maximized. Often getting less valuable but easy to implement features out the door is a great way to deliver value quickly.

If value is being scored the organization can even look at how the delivery of the customer value has changed customer prospection (via survey results) over time which can provide extremely valuable insight.

Key Take-Aways

- The Product Owners main goal is to create customer value

- Many organizations have a little contact and no way to collect feedback from customers that regularly use their products

- They will be multiple different types of customers using your products and each needs to be understood create customer persona so your team can quickly identify each type by name

- There are formal survey types such as NPS, CSAT, and CES which will allow much richer understanding of your customers wants and how they value your product

- Create a metrics scoring the customer value versus the difficulty/cost to implement allows for intelligent presentation of future work (a product roadmap)

Get used to saying no, and I don't know

To be a good Product Owner, you will need to get comfortable saying, "no" and "I don't know".

You want to be working on and delivering as much value as you can (do it fast, do it right, do it well), but there are limits, and it's better not to take on work than to break commitments to provide something you have already promised. There is only so much that your Development Team will be able to complete without sacrificing quality or burning out. No matter how well you are doing or how much the team is getting done, there will always be more requests than there will be the ability for the team to deliver.

If the team always says, "Yes, we will get that done," customers and stakeholders get used to you accepting all new work requests. They will invariably start sending you more work because someone else is being realistic about what their team can take on. It is far better not to take on work than to attempt work and later tell them that the team couldn't complete it. Or worse, you delivered it buggy or unstable and negatively impacted the reputation of your product with your customers.

What happens when your team gets overworked?

I came face to face with that when I was consulting at a top e-commerce company based in Seattle. On the way to my desk, our division's admin was boxing up all of the personal effects of a very sharp and critically crucial senior developer. It was very unusual to see this, so I asked, "Is Roberts moving to a new team?"

"Oh no," the admin answered. "He went to lunch three weeks ago, and no one has heard from him since. He doesn't return phone calls or emails, and he has run out of PTO, so they gave up on him".

Roberts was a talented developer, and his team counted on him, but he burned out and walked away from the company. Even with the best recruiters and extensive bonus checks, it takes months to recruit and hire new developers and months more before they are fully productive.

Losing top-tier talent never looks good and will always dramatically slow down your team's productivity. Worse than needing to replace someone was just how demoralizing this was for the team that he left. Everyone that he worked with knows that Roberts walked away from the company because he hadn't been able to change the unrealistic expectations, burn out culture, and he had no confidence that he was going to be able to change it. Everyone that was on Robert's old team was able to see that even after such a dramatic exit, that nothing changed, so many of them had to ask themselves "is this worth it", and "how much more of this can I take?". Is that really the type of work environment that you want to be fostering?

It's better to say no sooner to requests and give your team the chance to work at a realistic, sustainable pace than risk burning them out by taking on more work than they can realistically finish while still delivering a quality product.

Key Take-Aways

- There will always be more work than what your team can deliver

- It's better to say "No, we don't have the capacity" than to deliver low-quality or unstable functionality

- Developers leave when teams are asked to work at an unsustainable pace consistently

- You need to protect your team

Sell your technical debt

The business side of corporations is often overly focused on providing value and the latest whizz-bang feature and often ignores all of the infrastructures that deliver it to the end customer. This can be a very dangerous mindset in the long term if there isn't someone to correct that approach. As The Business representative embedded within the team, a large part of the Product Owner's role is to understand and prevent short-term thinking from sinking the product.

When I have a few minutes to spare, I often grab my iPhone and open up a game to unplug from work and relax a little. A massive studio with deep pockets produces one game that I like. The game received excellent ratings and was recommended by some well-known game reviewers a couple of years ago. It is a lot of fun to play, but the game has a big problem.

It crashes a lot, which kills the game's enjoyment. It's a head-to-head game where you play strategically against a real-life opponent somewhere else in the world. Like most games, you will gain rewards and access to better items if you do well enough. If a match is going poorly, some opponents will simply quit, and after a timeout has expired, you win the game. I've had an opponent down and, on the ropes, and the game crashed. It's frustrating to work hard to build up my play to be in that position, and because of the crash, a win was taken from me, as well as having a loss tacked on to my name that I didn't deserve.

The game has many players, a large following, and supports many in-game purchases, but there haven't been any updates for a long time and nothing that would fix this issue. The usual steady stream of updates, new characters, and new equipment has completely stopped, and the company has moved on to work on a new product.

I'm guessing here, but it looks like the stability issues have built up enough that it has become too difficult to track down the source of the problem and fix it. The product is bleeding existing customers fast enough that it makes more sense to put resources into building a new game than to improve what had been a very well-regarded top-tier game. It's sad because it is a delightful game, but that's what short-term thinking and technical debt get you.

Once you're delivering product in a semi-reliable manner, The Business tends to ignore that accomplishment, assuming that it's always going to be the case, and moves the focus onto the latest and greatest features. As a Product Owner, it's important to remember that the ability to deliver your product is more important than any given feature. Do you care how great your car stereo sounds when your car won't start?

Technical debt can be an enormous challenge. It isn't sexy, and it can be difficult to see. To keep it on the priority list and not have undue pushback from your Business Units, you will need to put in some work to sell the idea and value of fixing these issues. To do this, you will want to make it a regular part of your demos and give your team, and developers praise when major tech debt items or investments are completed. These fixes aren't easily demonstrated, but you can talk through the pain and frustration customers experienced and how much better the product will work with the new fixes.

For example, here are two ways of representing that same fix:

Fixed a stack overflow bug.

Or

"Our developers dug into a difficult issue which was causing user sessions to crash when more than one configuration was saved. This problem was painful for our most valuable customers, who always had multiple configurations.

The fix was valuable to a high-value customer and supported our product's strong technical reputation. Thank you, Neeha and Steve, for this excellent work."

Key Take-Aways

- How your product is reliably delivered is as important as all of the cool new features you could be working on

- Don't let a short-term feature-driven focus prevent the need to deliver your product

- Sell the idea and benefits of tech debt and infrastructure work your team does to ensure it gets the support this type of work needs

- Technical debt can build up dramatically slowing down future development of new features

- Explain how the fixes you make will benefit the end-users

Budget to ensure technical investments

A great way to ensure your product will be stable long-term is to develop a budget for the number of story points spent on features and the story points used to invest in your product.

Examples of product investments include; improving stability, logging for easier issue tracking, automated tests used to easily and quickly validate new product builds, and enhancing tools and test systems used to roll out new builds. Having a budget will allow you to keep an eye on your investments and maintain the right balance.

As for what makes sense, that will depend on where in the product life cycle you are. I like to use a base number of sixty percent features and forty percent infrastructure and then adjust up or down based on the product's current health. If you have stability issues, you should increase the percentage to address those issues. If you are working on an early-stage product that needs to get out into the marketplace, then moving to seventy-five to eighty percent feature-focused would probably be a smart move.

If The Business or Executives closely oversee your work backlog and priority choices, you will want to communicate and share your feature vs. technical investment budget. A budget provides the transparency needed and allows for intelligent adjustments based on your business cycle. It also shows that you are well aware of the stability and quality issues and plan to address them.

Regardless of how well the code is written, there will always be some crashes and customer impacts, so it is best to have a solid plan for dealing with them.

Key Take-Aways

- Creating a budget for how much effort your team will work on technical debt is a great way to ensure it doesn't get neglected

- There will always be some bugs found in production. Having a budget shows that you are not ignoring technical debt or quality

---

PART 3 -SHORT VS. LONG-TERM THINKING - BUILDING IN MAINTAINABILITY

The exception to the rule

Once a product is in front of customers, the Development Teams that created it will support it and fix any found issues. As a product adds new features and becomes more complex, it will require more time and effort to identify the source cause and resolve issues. As a Product Owner, you must work diligently to show the effort needed for this work, make sure that the effort required is quantified, and keep a regular drumbeat of good news about the fixed issues.

This seems like a minor task and could be easy to forget, but if the cost isn't visible, The Business could incorrectly assume that this work doesn't need to be supported and paid for.

Key Take-Aways

- Supporting a product in production will take effort from your team

- Make sure you build in capacity for production support work

- Advertise the product support work you have done at Sprint Demonstrations to keep visibility

Bigger isn't better

Clearly written and customer-focused User Stories are the tool your team uses to understand what needs to be delivered. As a Product Owner, you don't want your team wasting time fixing something that was just built when it could have been done right the first time with a clear User Story. Except for minor tasks, when the need is obvious, and the job won't take much time, you always want to have a User Story for your team's work.

When you get into larger projects or features, the question becomes how significant a User Story should be. Should I create one big story for a given feature? Or are multiple smaller stories better?

It's helpful to understand the drawbacks of any given approach so let us talk through the downsides of having overly large User Stories.

When User Stories are too large, they become much less likely to be finished during a given Sprint. Even if it looks like the team should finish it, any slight bump in the road can prevent a significant story from being completed. If you can't finish the work during the Sprint, you won't have anything to show during the Sprint Demonstration, and you can't claim the story points at the end of the Sprint.

Yes, you can save a little time by creating a smaller number of more meaningful stories, but that doesn't change the work that needs to get done. It simply lumps it into bigger buckets. When lumping all the work together in a larger story, more things must be completed before the story is done.

What if you get to the end of a Sprint and ninety-five percent of the work has been done? Claiming credit for finishing the story doesn't make sense because the story truly hasn't been completed, and accepting partial work for full credit is a slippery slope. It's also demoralizing for a team to work hard to get a story done only to end the Sprint with no credit because one or two tasks are unfinished.

Ending the Sprint with a huge incomplete story doesn't communicate to your stakeholders where the team made progress and what is left to be completed. If your stakeholders come to the Sprint Demo, you can explain where the team made progress and what still needs to be completed. But it's easy to forget the details, and if the stakeholder doesn't show up, you are out of luck.

Having too many things lumped into an overly large User Story also means that to get it done, the team will likely need to have multiple people working on it at any given time. Team members collaborating is productive and healthy, but only if they work together on a focused area. If the only commonality is the high-level goal, having multiple people working on it makes communication harder.

For instance, let's say you are out of milk, and your child will be grumpy in the morning if they don't have milk for their cereal. You could create a single User Story about going to the store, buying a gallon of milk, and getting home to get it in the fridge.

By using a single User Story, you need to include too many significant items such as grabbing keys and wallet/purse, putting on shoes before leaving the house, selecting which store to go to, driving to the store, and navigating the parking lot without running over someone's child.

Key Take-Aways

- When a User Story is too large, it is unlikely to be finished in a single Sprint

- Accepting a User Story as done when it hasn't been completed creates unhealthy habits for your team

- When a User Story is too large, it will require the work of multiple people to finish it in a given Sprint, which requires more communication and coordination of efforts

---

The motivation of the midnight phone call

Many organizations migrate slowly over to Agile, with many never making it very far. These halfway implementations will often keep the most convenient and often destructive habits. One of the most tempting is to have Production Support Development Teams (usually called Sustaining). The logic behind having these teams is that your best and brightest forge ahead, building extraordinary functionality without being distracted by the tedium of dealing with bug fixes. While this is tempting, the reality of what happens creates significant problems for the company.

The challenge of having large Production Support Development Teams - The primary Product Development Teams are under constant pressure to produce enhancements and new products, leaving someone else (anyone else) to deal with grumpy customers and late-night calls when things break. When someone else has to deal with the consequences, the quality and stability suffer because there isn't as much motivation to build a stable product.

The other problem is that the best and brightest team members move over to the Product Development Teams. The Sustaining ends up being a C-level team

(of course, there are some talented people, but if you average it out, they don't have the same horsepower). Not only does the team as a whole have less talent, but they may not have been involved in building the product in the first place, so they aren't as familiar with how it works, nor the trade-offs or weak points.

Of course, the Product Development Team is busy working on the next release and then the next project, so they don't have time to document things well and take time for a proper turnover. This leads to a knowledge gap for the sustaining team, or unexpected delays when the Product Development team has to stop their work to go back and explain how the code works to the sustaining team that is trying to fix critical bugs.

The communication gaps and lack of a quality mindset mean that when you have a separate Sustaining Team, they generally lack information and work with a lower quality product. Ultimately, the customers suffer, which drags down your product's reputation and stability and erodes profitability. This often leads to the death spiral where new features are rushed out the door without the quality that they need to make up for customer who are unhappy about the poor quality.

A better approach - When teams and management know they will be the ones to get the phone call if something critical breaks in production, calculations become different. Developers don't want to have their weekend interrupted, and they really don't want to be on a phone at three am in the morning trying to fix a critical issue that has brought down the website so they become motivated to build in quality and the ability to diagnose problems.

I'm not saying that making this transition will be easy because if the quality is low currently, it will be painful. But imagine what your customers are going through. Customers want reliable products and if you don't deliver someone else will so it's best to step up and fix the quality of your product.

If you are bringing responsibility back into the team,

I recommend having a good turnover with the Sustaining Team and spending a Sprint (and maybe more) to go through and harden/fix any of the broader issues impacting customers currently. You'll need to spend some time to do this, but having a quality product is essential, allowing your Development Team to own the stability of the product going forward.

Key Take-Aways

- When someone else is going to deal with late-night calls, there isn't as much motivation to build in quality

- Having a separate group to support production leads to an information and understanding gap because they weren't the ones who made the product

- For the best quality, bring production support into the team that makes the product. It takes more time and effort to build in quality, but your customers will be happier in the long run

---

Keeping quality in the team

As organizations take steps towards Agile, they may have a separate QA Team acting as a single point of contact for all quality work within an organization. While this may be convenient for people management, let's look at how that works when it comes to quality.

When the work is done in one group and then thrown over the wall to another (the QA group), that gap creates a significant communication barrier. Your QA Engineers won't have a complete view of how products or features are being put together. What information they receive is later in the product cycle, so they will have much less time to use the information.

By being on a different team, QA isn't included in the design discussions, trade-off decisions, and the context that leads up to the end product, making them much less prepared to do an excellent job of testing it. Having QA on a separate team also means that they can interpret the Acceptance Criteria differently leading to unneeded conflicts between the Developers and QA.

Along with a communication gap, people on different teams may feel their motivations aren't aligned. If I'm a developer who takes pride in their code, I probably won't take the time to help someone on my team who will tear in and try to break my brand-new feature. If they are on my team, I'd probably help them, but if they are on another team, I might be tempted just to provide minimal communication and focus on my next task.

QA engineers who feel like they are being purposely left in the dark can get bitter and further inflame the relationship between the teams.

To be successful, you'll need to bring QA into the team. If it's really not possible, you'll want to do everything you can to pull those external people into your team process as soon as you can to try to eliminate the communication and motivational gaps.

Key Take-Aways

- When QA is a separate group, they don't get the understanding and background with the product like they would when involved from the beginning

- Even with the best of intentions, communication between groups hampers coordination

- When in separate groups, the motivations of individual team members or the entire group may not align, causing the product quality to suffer

Taking the wins

Building products or software is complicated, but doing it with speed, agility, and quality is incredibly difficult. Things will go wrong, and it will get a little messy, so it's essential to keep up the team's spirit.

Even if you're on a team that's struggling, there will be things going well, and you want to be able to capture those. If you don't, no one else will.

That's reality. Development Managers have lots of meetings. Product Owners meet with The Business and customers (hopefully), so as a PO, it's going to be up to you to notice.

Finding things going well doesn't do any good if you don't point them out. Pointing them out helps the team notice them too. Studies have shown that it takes seven positive comments to balance out the impact of one negative comment so if you want to build the confidence and pride of the team it's going to take some work.

Positive comments give a boost to the people who are responsible, helps model positive behavior, and lets everyone know that you're paying attention and may share if you see something positive they've done.

This is especially important for struggling teams because everyone knows the team is struggling at some level, and it's never fun to work on flailing teams. Feeding your team is like feeding a pet. You could fill up their dish with a week's worth of food on Monday morning, but if you don't fill it again by Friday, there isn't going to be much left, and if there is anything left, it's going to be stale. It's best to provide small wins regularly. Of course, these need to be genuine. The last thing you want to do is lose credibility by trying to take a win with something that everyone knows wasn't a win.

Key Take-Aways

- Publicly recognizing and praising is a crucial way to model a healthy team culture

- Praising positive behavior helps motivate the team and encourages more positive behavior

- If your team is struggling, it's critical that you find good things to highlight

- As a Product Owner, you are uniquely positioned to see and provide positive feedback

---

Setting and keeping product standards

Setting expectations lets everyone know what to expect and what standards your team keeps. If you're new to the team or haven't reviewed it recently, going back over your standards documents is a valuable exercise and something you should do at least every six months. Without standards in place, team members may have different ideas about what "done" should look like, leading to confusion and conflict when one person considers something done but another doesn't.

The Definition of Done is one of the primary must-have documents for your team. The Definition of Done (DOD) is a generic set of Acceptance Criteria applied to every User Story. There will be times when something on the DOD doesn't apply to a story (like a research story that doesn't produce code won't be checked into source control), but if it applies most of the time, go ahead and include it.

For instance, the code for any story that has been completed should have a code review. It may seem obvious once you think about it, but some things aren't apparent (especially when you are trying to move fast), so it's best to spell them out and capture them. When the team is grooming stories, your team will concentrate on the Acceptance Criteria and not reinvent the wheel every time they write a story.

To complete a story, the team will meet the Definition of Done, which provides the foundation. Items unique to a given story are spelled out in the Acceptance Criteria. It works best to have the DOD copied into your User Story template, so there is a single location listing everything needing to be held to and accomplished for a given story to be completed (DOD).

The Definition of Ready is also an important document that spells out what needs to be in place before the team considers a User Story. If your team gets stories from someone other you, as PO, they must use the Definition of Ready. I've seen too many groups thrown into chaos by hot requests at the last minute that don't have enough information for the team to succeed. In these cases, the DOR can do a great job of shielding the team.

Stand firm and require that all new features and stories meet the Definition of Ready.

If you are getting stories directly from marketing or The Business, the urgent need to get these requests to you means that you often won't have the details needed. Sure, the team could make assumptions, but I guarantee that if you do, you'll have unhappy customers when the team guesses wrong. The development team will also be demoralized when their extra efforts aren't appreciated.

The best way to succeed is to have a firm, well-published Definition of Ready and practice the following: "I can see that this is important to you. This DOR details the information we need in order to do this work. Just give us these details, or sit down with us and we can help you with the information that we need, and we'd love to try to get it into the next Sprint."

You must train your Business to provide what you need. If your customer can't spend thirty minutes giving you the details, they either didn't need it that badly or hadn't thought through what they wanted.

Either way, you're not going to be successful until you get enough information.

Without the Definition of Ready, your team will waste time trying to guess what your customer wants, and that's going to make everyone on your team and your customer frustrated. As the Product Owner, use the above phrase; it will do wonders for your team's productivity and quality.

Key Take-Aways

- Standards are an essential tool in establishing the expectations and quality of the product

- The quality of the work and your team's reputation will suffer if people have different expectations about what is being delivered

- The Definition of Done DOD) is a list of generic Acceptance Criteria

- Most but not all of the individual items on the Definition of Done will apply to any given User Story

- The Definition of Ready (DOR) keeps teams from wasting time on stories that aren't ready to be worked

- Having a Definition of Ready shields, the team from customers who want things but won't spend enough time to tell you what they want

- The Definition of Done prevents wasting time and effort working on requests that don't have enough details

- The Definition of Done also protects the team from being blamed for delivering features that don't do what the customer needs

- Trying to guess what your customer wants is a losing battle and one you want to avoid

Develop a thick skin

Being successful as a Product Owner means having a thick skin and emotional intelligence to handle difficult situations. As a Product Owner you must be able navigate mad, annoyed, or grumpy people without destroying the relationship. In order to improve the team, you will need to be able to model constructive behavior and patience.

The Technology field is full of brilliant engineers who aren't used to dealing with people and often don't have strong people skills. It's also a stressful, fast-paced environment where the burn rate for a good Development Team could be six million dollars a year, and millions, and sometimes billions of dollars in revenue, are at stake.

Your role as Product Owner is to be the calm at the eye of the storm. You need to be the person to pull things back when emotions are high and to make peace when teammates are arguing over the product's technical direction. You're going to need to have a cool head, and when things get tense, here are some of the tools, patterns, and messages to consider/draw on.

Common Cause

Use when - There are disagreements on the team that might be getting personal.

Message - We're not always going to agree, and having different opinions is healthy because a diversity of ideas makes for better solutions. Discussion helps us understand the choices better, and come to an agreement or a third option that gives us more advantages than either of these choices. As a team, we all want a good outcome. We may not all agree on the direction that we will eventually take but we are all going to support it. We just need to talk through the pros and cons of the choices in front of us.

A Fresh Look

Use when - The conversation has progressed but has taken a different direction and is no longer working toward the original goal.

Use when - You think the solution or direction being proposed doesn't make sense or may not be considering major important factors.

Message: - Let's step back and review what we know so far (you walk through the information and significant points detailed so far).

A New Voice

Use when - The conversation is now stuck and repeating or not making forward progress towards the goal.

Message - "Thanks Dan, that's a great point. Does anyone else have thoughts about what could be happening in this area?"

Re-Center On Our Goal

Use when - The conversation has switched and is no longer progressing towards the original/most important goal.

Use when - Emotions are running high.

Message - "Ok, let's review; Can someone restate the end result that we are looking for?"

Let's Come Back To This

Use when - The conversation has gotten heated.

Key Take-Aways

- As Product Owner, a vital part of the job is to be the cool head in the room

- When tensions are high, some patterns can help focus emotions and energy away from interpersonal conflict and a fix

- Conversations can get stuck, and going over the same arguments is frustrating for everyone, so use patterns to get unstuck and move on

---

Shielding the team - being a shit umbrella

A little understood but critical role for the Product Owner is the need to shield the Development Team. When they are allowed to focus, good Development Teams can create tremendous value and emotional capital. Of course, if enough distractions happen the team may become completely unproductive.

In the early days of my career as a Scrum Master at a large bank, I thought I should keep the team informed of everything I knew or found out about the product's future direction we were working on. During Sprint Planning, one week, a senior developer looked at me and said, "Why should I work on this? The Business has canceled and restarted this project four times already. If I start working on this, they are probably just going to cancel it again in two weeks, and I'll have wasted all of my time."

Right there, I realized my mistake. Instead of giving the team every little bit of product gossip I'd been collecting, I should have only brought them the work that had been committed and detailed.

Key Take-Aways

- The Product Owner and Scrum Master need to shield the Development Team from distractions so they can focus on coding

- Not everyone in your company will prioritize the work the same way

How to eliminate a major distraction for your team

In large companies and corporations (especially those with a more formal business process), it is prevalent to have an informal or underground economy. I'm not talking about selling hot Honda parts to beef up street racers, rather about getting around the normal work request processes in place to vet and approve new items.

There can only be one thing at the top of the priority list, and everyone else would love to find a way to get their priorities done, even if it means using alternate and unofficial routes. Getting a new feature rolled out the door to stay ahead of the company's competition might be the highest priority for the company.

However, a customer support manager who has a massive bonus on the line and a customer who has been complaining about a bug in a seldom-used feature might see things differently.

If back channels and work on the side exist, they will grow and detract from the team, resulting in lower quality, lower velocity, missed deadlines, and frustration within the team because they can't get everything done. It could make sense stop the current work to find and fix a new hot request, or maybe find the fix if it doesn't take up too much time, but it should be a conscious decision weighed by the Product Owner and The Business, not a competition for who can find new ways to slip work into the development queue.

Deciding how and why million-dollar budgets are spent and how the company should pivot its applications and custom tools requires much debate and negotiations. There can be a lot of politics and drama around who and how strategic direction is decided in large organizations.

This is distracting and de-motivating for Development Teams.

Whenever someone gets around the official process and gets something done in half the expected time, the more this encourages them (and others) to try again, and the more likely they will get a yes from someone on your team who has already helped them out in the past. After repeated success, the new workaround becomes the default way of doing work, and the checks put in place to ensure that your team is working on the right things get bypassed. The team will have less time to get critical work done because they are busy doing work for someone else.

If side work is accepted in this fashion it will also invalidate all of the estimating and commitments that the team has done, likely making the team look bad because they can't deliver on commitments that they have made.

It's possible that some (or even all) of the requests coming in are important, and the team would end up doing the work at some point anyway, but this doesn't negate the problem. You must understand the underground economy, the costs to you and the team, and if the new requests are indeed more important.

As long as you ask carefully, in a casual, non-judgmental way, it's easy to uncover and find out if your team is being affected by the underground economy. Simply add the following at the end of your next Standup, "Hey, I'm just curious… Is anyone helping out or doing any little tasks to support other teams?"

You may need to ask weekly for several weeks, but as long as the team doesn't think you'll be upset (and the team has Psychological Safety), you should be able to uncover what type of requests are being asked of your team. Finding out about it is only the first step, however, as you need to understand how much work this represents and its impact on your team.

Next, talk with the people who are fielding these requests to understand how often they are coming in, what type of work is being requested, who is submitting the work using this back-door approach, and why. The requests may be small, require little time, and are helpful to the organization as a whole, but they are, in fact, a problem.

It may be difficult for someone in the habit of helping with these requests to cut them off abruptly, "No, I'm not going to do that," because they did it last week. The most effective way to combat the underground economy is to explain to your team member that some of these requests may not make sense to do, and you want to protect the development team. But you need to have visibility into the requests, so they should be routed through you or the Scrum Master.

Teach your team that they don't have to say no and can instead offer the following:

"The request makes sense; I can see why you need that; just route it through my Product Owner, so they know where I'm spending my time."

This approach eases potential tension and allows the team member to leave the person hopeful without committing. If the Product Owner needs to say "No, the team member isn't "the bad guy," the PO is." Using the above phrase is more effective, more likely to get used, and it still routes the request back through someone who can adequately judge the priority of the work against the current workload and overall strategy.

If the request for side work is coming from an executive then the organization has larger problems but that doesn't mean that they should be ignored or just blindly delivered. Actually, requests from executives become even more important because you don't want to let them fall on the floor, and you definitely want to capture the level of effort being asked for. So, the best thing to do it to capture the request in User Stories, estimate the work, and build it into the team's capacity so that the team isn't blindly over committing.

Key Take-Aways

- Side work can be a major distraction for a team and ruin productivity

- Find out how much side work is an issue by asking your team in a kind and non-judgmental fashion

- It's easier for individual team members to redirect requests to the PO instead of having to say no

No untracked work

When I started at a new data visualization company, I supported an existing team. I did my homework by looking over the backlog and the User Stories to look for write-up quality and completeness, thus getting a background into their work focus.

During the next Sprint, I noticed a difference between the story points in the User Stories and how many story points were allocated for the team's velocity. I was told that the team allocated ten percent of their time (roughly eight-story points) for product support issues, so it wasn't necessary to track that work directly; it was built-in.

I checked in with a senior colleague to see what production support issues they typically encountered. I listened to a long list of complex and ongoing support that sounded like a significant time commitment, not a small side project.

I decided to test my theory and talked with the two developers who did the most production support work. I asked if they could keep a short one-line log to capture the production support tasks and story points they completed. At the end of the Sprint, I collected the lists, got last-minute updates, and tallied up our actual spend for production support.

The team was shocked when I shared that they spent thirty-seven percent of their time fixing production support issues, far greater than the ten percent reserved for this work. The team suddenly understood why it was difficult to deliver on their Sprint commitments.

I'm sure that reserving ten percent for bug fixes was a figure that made sense at one point, but over time as more and more functionality was released, more and more support was needed (there was a technical debt problem). The need to support production had slowly grown to have a huge over-sized impact. Problems can spring up quickly, and generically reserving a certain amount of effort won't give visibility into what is really happening.

A best practice is to teach the team to capture work items that take more than an hour. You can change the granularity of how you measure it, but anything that takes a significant amount of time should be accounted for. You don't want to ask developers to spend fifteen minutes creating a full formal User Story to track an hour's work of production troubleshooting. Keep tracking easy.

I like to use tracking stories. These are lightweight stories only used to track short duration and easy-to-understand work, and the only thing required is a title, a tag of "tracking story," and an estimate for what was spent when the effort is done.

You or the Scrum Master can easily collect these tracking stories at the end of the Sprint and detail how many story points were paid. Accuracy is essential, so it works best to have the team capture the effort after doing the work. You can't predict how long it will take with production support work.

Key Take-Aways

- You can't know how to manage your work until you can accurately understand where the work is being done

- When tracking work, the User Story (tracking story) doesn't need all of the details that it would have for feature work

- It is easy for maintenance work to grow over time to become more significant than you expect it to be

PART 4 - KEY IDEAS

How a Product Owner different from a Program Manager

The Product Owner and Program Manager are very different jobs with distinct goals and responsibilities. Let's take a look at the critical differences.

Program/Project Manager – have a temporary relationship to teams or products to tracks the execution and completion of a defined set of work. For example, they might follow software installation for a new customer or the progress of new features being built for a new or existing product. The work tracked is primarily defined by someone else, so the Program Manager's main focus is to keep track of a broad group of people or teams, the work each is doing, and the overall completion of the program or project.

Product Owner – responsible for the continued quality and value of a product or portion of a product. In this role, they prioritize and define the work needing to be done (User Stories and bug fixes) and accept the quality of the new features and bug fixes from the Development Team they work with.

In Agile, Product Owners work directly with the Development Teams, check the quality, and accept work from teams (completed User Stories and bug fixes). Product/Program Managers focus on stitching together the work from various teams to create the overall product.

Program/Product Managers are responsible but not accountable for the completeness and accuracy of the work defined for the Development Team, only the execution and tracking of it.

Product Owners need to have expertise and oversight over a particular product area because they define the work and own this area of the product on an ongoing basis.

Because their ownership of a product doesn't change or disappear after a release or milestone, the Product Owner needs to consider the long-term health of the product and not just make short-term decisions that might have dire long-term consequences.

Key Take-Aways

- The Product Owner role has different responsibilities and deliverables than the Program Manager role

- Program Managers track the completion of work.

- Temporary relationship to the team and the product

- Product Manager is a staff position accountable for the lifecycle of a product

- Product Owners are accountable for defining/creating value for a product or a portion of a product

- Product Owners are embedded within Development Teams

- Product Owners are responsible for the quality of their products

Keeping a balance

As a Product Owner, you will continually decide where your teams should be spending time, effort, and resources. You must keep the balance between new features and providing a reliable, stable product. When you are making good choices, these decisions will fall into the following categories. It's essential to keep a balance between them.

Build the right thing - put effort into creating new features your customers or your company will find valuable. New features and functionality can set your product apart, keep existing customers, and attract new ones so it's important to work on the right stuff and not features that won't be used.

Build it fast (speed) - focus on getting something done and out the door to gain feedback from the customer.

Building it well (quality and stability) - create something that will be maintainable, stable, and testable through automation, so it won't require significant time and effort to diagnose or fix in the future.

Investing in your team - build the skills and understanding within your team to allow them to work smarter, with better tools, and to learn approaches that will generate less technical debt, improve morale among your developers, and require less time to fix issues in the future.

Key Take-Aways

- There are four main areas you should focus on: building the right thing, speed, quality, and investing in your team

- It is essential to balance and not ignore any of these areas

The Product Owner's two deliverables

While there are many decisions and choices to be made to help the team deliver value, the Product Owner focuses on 1) creating User Stories (and features made up of User Stories) and 2) prioritizing User Stories.

User Stories are short, self-contained, customer-focused units of work describing the end result that an internal or external customer need. User Stories have a set narrative format which spells out what the end customer is hoping to accomplish and why that is valuable to them. User Stories should keep to a set format which is easy to understand using the word **INVEST**.

I - Independent: the story is as independent as possible

N - Negotiable: the story details are negotiated between the team and the PO

V - Valuable: the story is meant to capture meaningful valuable work

E - Estimable: the effort is detailed enough that it can be estimated

S - Small: the story can be completed in a sprint or less

T - Testable: the stories output can be validated

By prioritizing User Stories, the PO defines what needs to be created and in what order to best provide value for the customers.

As a product expert, the Product Owner represents The Business and the end customer's needs. They write the User Story and the Acceptance Criteria, which define what needs to be accomplished for the User Story to be complete. The User Story isn't ready to work until the Development Team has had a chance to review it (usually in a ceremony called "Backlog Grooming"), and have a conversation with the PO if they have any questions or need clarification.

Once the team believes that the User Story is completed, they will bring it back to the Product Owner, reviewing what the team has accomplished. If the Acceptance Criteria are met and the resulting work has the quality it needs, the PO will accept the story.

Let's look at the different pieces that make up a good User Story by examining a sample User Story template. This template effectively captures and communicates any details the team may need.

User Story Template

Title - Short, catchy, self-descriptive

Customer need (you may need more than one customer statement) **-**

- As a
- I want to
- So I can
- Which will

Acceptance criteria (what evidence/results should be seen if this story is complete) -

What evidence of success should exist?

What should not have been changed?

Assumptions -

What isn't included in this work?

What are we assuming will exist/happen?

Contacts -

Details -

Notes -

Team Definition of Done (if available) –

Key Take-Aways

- Following a good User Story template is an easy way to write good User Stories

- The User Story template won't fit every situation and should be customized as needed

- Creating and prioritizing User Stories are the only two deliverables that a Product Owner has

Quality pays off

Having quality detailed User Stories pays off because it allows the Development Team to understand what is needed, what needs to be accomplished for the story to be complete, and what the customers goal is. Quality User Stories dramatically reduce the need to rework and rebuild features, saving your team time and making a better end product.

But what happens when the User Story doesn't have enough detail in the beginning? I like to use this analogy as a way to show what happens.

Imagine that creating a car is something that your team can do in a Sprint. You have made a high-level User Story describing only necessary details about the vehicle you want your team to build. Without detailed Acceptance Criteria in the User Story, your team has a lot of flexibility in delivering the car, so they start constructing a new car. A small zippy two-seater sports car won't require much work and will be fun, so they decided to build a little BMW Z4 two-seater convertible.

Now halfway into the Sprint, you find out that this only carries two passengers, so you share with the team that this car will be used by a family of four, so it needs a backseat, and the current design doesn't have one. Oops! You add this detail to the Acceptance Criteria, hoping that it won't have much impact.

Your team now has to turn this half-completed two-seater sports car into a family sedan. The car's frame had already been built along with the powertrain, which presents a problem. Your team now has to cut the existing frame into two pieces, construct a backseat that fits between the two parts, and weld it all together again. The extra three feet added also means that the transmission no longer fits, and the drivetrain won't reach the back wheels to provide the power needed to turn the wheels, so this too has to be fixed. The small trunk won't work anymore because there will be more people and luggage in the car, so it will need to be removed, the frame adapted to a new design with more space, and a new trunk constructed and put into place.

It's all something your team can do, but in the end, it will take much longer than if the details had been in place from the beginning.

The other big challenge is that the car's frame has been built, cut into pieces, has new parts added, and then welded back together, so it's not going to provide a smooth, seamless design like your customers expect. And if it gets involved in a crash, the welds may not hold up like they would if the car's frame was constructed out of one piece like it should have been.

In the end, spending the extra time you need to write up a detailed and complete User Story is time well spent as it will save your team a lot of time and effort.

Key Take-Aways

- Incomplete User Stories lead to rework and lower quality

- Investing the time to write a good User Story will save your team a lot of time and effort

User Stories are like donuts

User Stories are essential, but they are like donuts, and you don't want to have too many of them lying around. Life is good when you have just enough donuts for you and your team, and they are fresh and up to date. No one wants a User Story that's old and stale, right?

When a User Story has gotten too old, the details and context around it have likely moved on. The Business often has new or more in-depth knowledge about the customer's needs, and the priorities that drove the story to be included may have changed as well. The architecture and effort to implement the story may also be different if it has been more than one Sprint.

Once a User Story is too old, it needs freshening up. The Product Owner must re-read and prioritize the story with fresh eyes. Then the story needs to be re-groomed by the Development Team. It will take less time to freshen up a story than to create it initially. Still, it's not uncommon to find that an old account is no longer needed or that the priority is so low that it won't make the cut for several Sprints (when it will need to be freshened up again).

Creating quality and relevant User Stories takes an investment of time and effort by the team and Product Owner, so it's best not to get too far ahead in writing them. Having enough User Stories for the upcoming Sprint plus another half a Sprint's worth of stories generally is about right. You'll have enough stories to keep the team busy, even if you need to add a story or two, but you won't be investing time into stories that will need to be reworked again later.

Key Take-Aways

- New details and priority changes that might change a User Story happen regularly

- User Stories get stale once they are too old

- Stale stories need to be re-read by the Product Owner and updated before teams can work on them

- Building out too many Sprints worth of backlog wastes time because stories will need to be refreshed

---

A Product Owner's one goal and one strategy

Your highest goal as a Product Owner is to demonstrate to your company, The Business, and your boss that your team provides innovative, focused, end-customer value while improving your product's quality and reputation. The best way to demonstrate this is to deliver an intelligent, focused demonstration of what your team accomplished at the end of every Sprint.

There is another great benefit of holding regular Sprint Demonstrations. During the rollout to production, many things can go wrong. Your secret weapon to avoiding production problems is to use the End of Sprint Demo to show off your team's work and practice rolling code out to production. The more like display your demo is, the less risk there will be when it is released to customers. The ideal is for the demo to be run on a production-like machine, with a fresh install of newly built and configured code.

When your product is first being created, it's going to be a challenge and likely impossible to be able to run your demo from a production-like environment.

Having code working from a Virtual Machine from under a developer's desk will not give you any reassurance that it's going to be working in production as it should. Hence, as Product Owner, it's up to you to push the team to make the demonstration as production-like as possible. You will need to push the team continually, and likely your executives, to improve and make the demo more production-like for each Sprint.

The ultimate goal for your demonstration is to have an untrained person from customer support or sales be able to install, use, demonstrate a working product, and new functionality with the same permissions, hardware, and configuration that your customers will be using. When the demonstration is run exactly how the product will be used in production, you will know what to expect before delivering it to customers.

Key Take-Aways

- As a Product Owner, your goal is to deliver value to your internal and external customers

- Sprint Demonstrations are a crucial tool to show the successful completion of customer value

- Setup your demonstration to be as production-like as possible

---

Don't ignore technical debt

Technical debt is when getting features out quickly has been prioritized over the long-term stability or quality of the product. It's called debt because, like financial debt, its impact grows over time, and if it isn't repaid (and repaired), it can have severe effects on the future of your product.

Technical debt is a lot like used gym socks. If there is a little around for a short time, it's understandable and tolerable (as long as it's hidden away in a hamper somewhere), but it is easy to build up because no one wants to deal with tech debt. Once tech debt builds up, it becomes a real problem, and many products or even companies have been sunk by too much technical debt.

If you're honest with yourself, technical debt exists because people or teams have cut corners to get features or releases out earlier than they should have been. It's a product of short-term thinking while ignoring the long-term costs. Some inexperienced Product Owners will drive teams to take whatever shortcuts possible to deliver functionality earlier but then won't want to go back to clean up the mess that's been created.

As more and more technical debt builds up, it requires more time and effort to build new features, and more and more time is spent patching and stabilizing the code that's in production, robbing the team of the time to build new features. A year ago, the Product Owner was delivering things artificially fast, and now with a room full of smelly technical debt, new features are being delivered artificially slow because so much effort is spent just trying to work around and stabilize the code.

To work smart, don't allow technical debt to build up in the first place. It's still critical to delivering new functionality, so don't rob your future self to try to look better today. It's best to set realistic expectations for how long it will take to deliver functionality in the first place.

Key Take-Aways

- Tech debt happens, but don't let it build up

- When tech debt builds up, it robs you of development time

The Cost of a poor reputation

As a Product Owner, you need to keep the balance between new features your customers value and keeping your company's reputation intact by providing a reliable and stable product. If you fall too far behind on useful features or product stability, your customers will leave and find an alternative. You might be working for a company fortunate enough not to have significant competitors in your current market. Even so, there are always steep prices to pay when your products have a low-quality reputation.

I won't turn up my nose at Microsoft's success, but they are a great example of what can go wrong when quality doesn't get enough focus. Microsoft makes billions of dollars and sells more every day than the entire economic output of many countries. However, it's a company plagued by quality issues and a low reputation with customers. One big positive for Microsoft is that they have built-in leverage because most companies use the Windows operating system and Microsoft Office Suite. Microsoft will continue to sell billions of dollars' worth of software to its corporate customers with a strong sales group. But its reputation with end-users has suffered mightily.

To see this cost in action, you only have to look at the current day usage of the most critical and valuable customer-facing software in the world today. The portal to the internet is the web browser. Let's look at the most up-to-date market share information ranked top to bottom.

Chrome – 64%
Safari – 19%
Firefox – 4%
Samsung Internet – 4%
Microsoft Edge – 3%
Opera – 2%
UC Browser – 1%
Microsoft Internet Explorer – 1%

Eleven years ago, Internet Explorer ranked higher than Chrome, with sixty-five percent of the web traffic on the internet. In 2010 Internet Explorer's market share dropped ten percent, and it's been shrinking ever since. Microsoft employs many intelligent people, and they recognized they had a problem. Fixes and new versions of Internet Explorer rolled out after 2010, and you can see how the end-users failed to adopt them and continued to migrate to alternatives.

Microsoft Edge was built from the ground up and was meant to be a fresh start with the hope of leaving the bad reputation and slow, bloated code base of Internet Explorer behind. After much fanfare, advertisements, and fast benchmarks, the end-users have rejected Microsoft's browsers leaving a once-mighty billion-dollar product to collect dust in the corner.

When a product gains a bad reputation for quality, it becomes challenging to cure the negative perception because new features won't entice customers when competitors offer comparable features.

You have trained your customers not to use your product, and using your product is the only way that you can prove that the problems have been fixed (if they have).

Key Take-Aways

- Neglecting the quality and stability of your product will ruin your product's reputation with customers over time

- Once a product has developed a low reputation, it is easy to lose customers and tough to win them back

---

What is the Standup for?

Most people learn that the standard model "What I did. What I will be doing. Any help needed." works, but the team is leaving value on the table if someone is trying to keep the Standup's as short and sparse as possible. You don't want a Standup that drags on with people droning on and on, but you will want to go beyond the basics to do the best you can for your team.

A good Standup works on several levels, and changing your team's mindset is the first step.

The first change I like to make is to change the name. Yes, the methodology is called Scrum, but do you want your team to lock arms and try to hook an oblong ball out of a pack of sweaty guys every morning? Probably not! It's best to change the name to what you want it to be, a "Daily Planning Session," not a sports reference.

If you give the word Scrum to ten different people, you'll get at least six different definitions, and you want everyone to be on the same page as to what it is and its purpose. I call it "The Daily Planning Session" because that is what you're hoping to do.

Here are some of overlapping goals of a good Daily Planning Session.

1. Find the best ways around the broken builds, tools that won't work, the information you don't have, or anything else that will allow your team to make the best progress possible

2. Help your team members understand what other people are working on to better understand how their work fits into the working whole

3. Provide visibility so other team members can make suggestions, provide help, and collaborate. "I ran into that problem last year, and I fixed it by..."

4. Identify adjustments that would allow the team as a whole to be more productive (an hour from someone who knows how something works will save 6 of someone who doesn't)

5. Over time, establish who is good at what

6. Over time, build the background knowledge of the less experienced team members

Guidelines for a good Daily Planning

1. Coming out of the meeting with a good plan for the team to be productive is more important than keeping a set timeline

2. The Standup should be focused and driven by the team, and the team should do most of the talking (not the Scrum Master or Product Owner)

3. Don't discourage sharing if it's gone on too long; acknowledge its importance, thank the sharer, and add it to the Post-Standup After Party

4. The After Party can be as crucial as the Standup, so stick around and support it

5. The parking lot is something you get stuck in; an "After Party" is a better name for the Post- Standup meeting

6. Provide team members a little slack if they veer off track; they'll usually refocus and finish

7. Acknowledge and commend functional behavior like collaboration, helping teammates, and sharing valuable information. After all, these are the reasons you have Standup

8. Provide a clear opening for the After Party

9. Keep a record of the parking lot items so they don't get lost

10. Set it up so if you have to leave after Standup, the After Party can continue as long as it needs to without you. (Zoom or Webex options)

Key Take-Aways

- Don't limit Standup to a set timeframe. Quality is more important

- Details about what team members are working on provide a way for less senior members to learn

- Details about what team members are working on let people understand how their work fits into the whole team's effort

- Information about the challenges or problems they are facing provides a chance for the team to adjust or get help

---

PART 5 - THE PRODUCT OWNER'S CORE CUSTOMERS

The Origin Story - Why was the Product Owner role created?

The beginning

Superhero themes have been a significant movie trend in recent years. Rich characters, backstories, conflict, and drama provide an endless palette for painting stories. For any given superhero or villain (and who doesn't love a good villain), the most important story is the origin story.

To understand Batman as a character, what drives him, where he derives his strength and why he does what he does so well, you need to look at where his story began.

Agile, as a discipline, barely acknowledges the Product Owner, seldom goes into any detail, and never examines the origins of how the role came into being. The Product Owner role didn't just spring into existence overnight, and it certainly wasn't dropped down a chimney, delivered by a stork, or found under a cabbage. So how did the Product Owner role begin? The Product Owner role was created using a set of best practices gathered from people who were doing a great job and working directly with the Development Teams.

In the traditional process, different groups with a stake in the work being done by the Development Team encountered the same problems and issues repeatedly.

The Product Owner role was created to solve the major issues that stakeholders continually encountered.

To help you understand the role, I'll go over the different stakeholders that interact with Product Development Teams and the issues typically encountered when using the standard Product Development Process.

If you have not experienced a more traditional Product Development Process (consider yourself lucky), I've included a quick background (which you can skip if you've already familiar with the process).

The background

The traditional product development cycle had a six to twelve-month schedule consisting of three overlapping phases. Based on the development phase, different teams and roles had more (or less) work.

Phase 1 - was to gather The Business's details about what the customers and The Business needed; multiple groups and people prioritized those and captured them in lengthy, written requirements documentation, typically taking six to eight weeks to complete.

Phase 2 - started when requirements documents were given to the Development Teams, who then began to understand and start working on the list of features and fixes needed by The Business.

Program Managers tracked the progress of the efforts and dependencies by regularly checking with teams and individual developers to see what percentage of a given item had been completed on a given day and how much was left to do. As Program Managers collected status, the progress of the release and its ability to meet or not meet the scheduled release date was tracked using complicated timeline-based schedules.

Phase 3 - The Product Release schedules created at the start of the project were rarely kept, but once the Product was deemed ready for release, it would get final testing and then be rolled out to the customer (who had been waiting a long time to get the improved version).

Key Take-Aways

- The Product Owner role was created from a set of well-established best practices that came into being to solve weak spots in the traditional product development approach

- To understand the Product Owner role, it is essential to understand the background and inefficiencies the position was designed to solve

The Executive Leadership

A key stakeholder group are the executives that oversee the Product Development programs. Being in Executive Leadership for a Product Development program using the traditional product development lifecycle means that you have problems coming to you from all sides. The Business will let you know if something goes wrong. The Development Teams are hard to staff, expensive, and have challenges getting details about what The Business needs, and feedback on completed work before it gets released.

The Business complained to Executive Leadership when the quality or completeness delivered wasn't to their expectations. There is bound to be something they are unhappy about with considerable risk and complex software releases.

An enormous difficulty for the Executive Leadership is that many of the Development Team's challenges don't come to light until late in the traditional Product Development Life Cycle. Because these surprises were late in the cycle, they severely limited mitigation options. More uncomfortable was how having these high-impact surprises called into question the competence and value of the Product Development's Executive Leadership. I saw this in action when I worked at a medium sized company that built software for the publishing industry.

The updated software was scheduled to release on May10th but didn't get released until July 1st. A week before the software was scheduled to be released the Development Team was scrambling to fix a feature that the Business had deemed to be user hostile (even though it had been built based on the Businesses design), and another feature that had fallen through the cracks. After the release on top of bugs being encountered by unhappy customers two features promised to important customers weren't in the release while multiple other small priority items were. There was a lot of finger pointing between Product Development, and the Business, but ultimately it was the Executive Leadership that looked incompetent.

The Product Owner's Role for the Executive Leadership - For the executive leadership, the Product Owner is the most critical view of the teams' work and efforts.

The PO serves the following roles for the Executive Leadership.

A single point of contact to understand progress - The Executive Leadership needs to have an up-to-date understanding of the progress being made within the Development Teams and any business challenges. The Product Owner is a single central point of contact to get both.

An inside view into teams - While the Scrum Master provides an inside view of the team dynamics and process, the Product Owner provides a crucial second viewpoint into how the team functions, what to improve, and what's working well from a product development perspective. This second viewpoint offers important visibility for the Executive Leadership, enabling them to adjust and improve the output and stability of product development.

They didn't know about this before now - Getting warnings about potential issues is critical as an Executive Leader. With enough notice, mitigations can be put in place. Leaders can formulate how to message any issues encountered, demonstrate their value to the organization, and improve the output of product development.

You're in the same department, didn't you talk - As the complexity of products has increased, the need to coordinate across Product Development Teams has increased. As someone who fully understands the work being done and how it fits into the strategic priority, the Product Owner is uniquely qualified to coordinate across Development Teams. Executive Leadership relies on the PO to fill in and conduct the day-to-day coordination they can't.

There are ways of doing things here - As the person who defines the Acceptance Criteria for all User Stories and then accepts or rejects the team's work, the Product Owner is the primary means to support the standards and best practices of the program. Often some of the product level standards are defined by the Executive Leadership. The best way to uphold these standards is to have the commonly used ones built into the Definition of Done, which acts as a generic Acceptance Criteria which must be met for all User Stories.

Standards that are more specific and apply only occasionally should be kept in a list and added to User Stories Acceptance Criteria. Before accepting User Stories as being complete, the Product Owner upholds the standards set by the Executive

Leadership by reviewing the completeness of the Definition of Done and Acceptance Criteria.

Not like that, like this - As a QA Manager for a startup dot-com company in Seattle, working with the publishing industry. The first release, that was scheduled to go out in six months, turned into a slogging death march. A significant part of why it went so long was a very opinionated and self-assured Business Analyst, whom I'll name Sarah. After getting the latest view into a new feature, Sarah storms over to the Development Manager to complain,

"John, I thought you said you were going to fix that menu!?"

"I did; we changed it to be exactly how you requested it," John explained.

"That's not how I asked you to build it; look at this, it's going to be confusing for the customer, we need a row of buttons..."

and off she would go into a new version of something useful but not the style that she wanted on that day. The trouble was that she had selected many of the original choices. Over time some items had flip-flopped two or three times. In the end, it cost the company a lot of time and money and didn't improve the product or the customer experience.

Another major challenge encountered was having data structured in a specific way. What The Business wanted could be easily accomplished in a customer-friendly manner with two weeks' worth of work, instead of the twelve weeks it would take to achieve it using the exact solution The Business requested.

Being micromanaged is never fun, but it can be excruciating for a Product Development Team. Creating a working and functioning product or feature is challenging enough. But when you have someone swooping in to tell you to change and re-change a feature, it can feel like you're Sisyphus, pushing the same big rocks up the same big hill every day. This management style is sometimes called "Seagull Management" because the Manager swoops in, poops on things, and then swoops out.

An important but sometimes overlooked approach for Agile is to remove The Business from design decisions and empower the team to develop suitable solutions without being micromanaged. Doing so gives the team the ability to work brilliantly, feel empowered, have ownership, and use an approach best suited for the situation. To be successful organizations need to hire smart capable people and if you don't empower them, they are going to find someplace else to work.

Protecting your team's autonomy and ownership of the solution is a critical part of your role as a Product Owner. Particularly at companies that have recently switched to Agile, The Business is used to dictating what the solution will look like and micro-managing the end solutions.

For your team's productivity and autonomy, it's important to ensure that appropriate design decisions are kept within the team.

Key Take-Aways

- Being micromanaged is demoralizing, and good developers will often find a new job rather than put up with it

- Developers who understand how the data is structured and the existing code can find better and cheaper solutions than anyone else

- As a Product Owner, it is crucial to protect your teams

Challenges for the Development Team

I remember calling a meeting to clarify the priorities for an upcoming release. I walked into the conference room, got set up, and five minutes late, the business guy rushed in, shook a few hands, and announced,

"I'm not sure why we need this meeting in the first place."

"Because we need to know the priorities for the upcoming release," I said.

"Yes, and I sent you those on Friday," he declared.

"Yes, yes you did, but there are seven number one priorities. So, which is the number one priority?" I asked.

"Well… they all are. I need all of them!", he explained.

"Yes, I can understand that, and I think we'll be able to complete the list you sent, but which out of seven are the highest priority?" I pressed.

"They are all the highest priority!" he said.

It was time to switch tactics, so I asked, "If you could only pick one thing to complete this release, what would it be?"

"Oh, we have to have number two and number five. We have to have both of those".

In traditional Product Development, working on a Development Team could be frustrating. Information was needed to prioritize work and deliver the features and functionality as expected, but it was seldom received from The Business. It wasn't uncommon, after months of hard work and two or three weeks of long nights and weekends, to have the release roll out to a resounding thud, and the team would get chastised for what they had delivered. You don't have a team of mind readers.

Not enough - I made a rookie mistake that I'll never forget in my first job as a Program Manager (before I started using Agile). It was a painful lesson, but most of the great learning opportunities are. The company was using a six-month software release cycle. We were four and a half months through and weeks away from closing out the new feature development portion. It was time to focus on intensive testing and bug fixes as we marched towards an end release to the customer.

Sitting down to a sourdough sandwich, I began to read through my email. I saw a meeting invite from someone I didn't know from a group that supports the company's most important clients. After a brief discussion, the Sales Manager explained why getting a significant new change into the current release was necessary for a high-profile customer. It was going to be a stretch because there wasn't much time, and there was only sparse information about what was needed, but I was new at the role and wanted to show that I could be flexible and deliver value, so I told him that I thought I could get it in.

It was a Friday, and there was only one week of development left, so I needed to get details from him on Monday so work could start right away. The Sales Manager promised I would have all of the details I needed on Monday, and we both went our separate ways.

Monday came and went but I never head from that Sales Manager. By the end of day Monday, I had left a voicemail, a chat, and an email asking for more information. Tuesday morning, still dead air.

The window to get the feature work done was closing fast, so I took matters into my own hands. I was a brilliant, capable technologist, wasn't I? Shouldn't I be able to fill in the gaps and lay out what was needed to help the end customer?

I walked out into Pioneer Square, and splurging a little, I ordered a bento box lunch, sat with my laptop under a heavy, gnarled oak tree in a brick square, and worked on detailing what was needed. After a long lunch, I had worked my way through a fantastic piece of chicken katsu, and I had a good draft for what was required. I took it back to my team, and they got to work.

After work had started, one of my senior developers pulled me aside, and a little skeptically, asked if The Business guy had come back with the details we went over. I explained that he hadn't, but it was straightforward, and I had figured out what we needed to do. I never did hear back from the Sales Manager, so I didn't make any adjustments to the original design I created.

When we released, I was proud that our team had delivered, but a day after we had, I got news from my boss that the Sales Manager had been complaining that what we delivered wasn't what he had asked for or what the customer needed. I was angry; the Sales Manager had never given me the details required, but I didn't push back because I was the one that failed.

Given the circumstances, I made the wrong choice by putting my team in that position. The team should never have been working on customer-facing features without details from the customer (or a customer representative) about what was needed. I have learned my lesson. If someone can't provide enough details about what is needed by the customer, it doesn't make sense to spend resources to build something and hope it will deliver what is needed.

Locked into a feature - While not having enough details is a serious problem, some teams have the opposite problem. When working at a prominent retail stockbroker, I worked on a team with one customer that went a little overboard. They would provide in-depth details about what was needed (excellent) but then handed over pages and exact specs on how everything would work. They never consulted with the team before constructing all of this, and often what they were asking for couldn't be done with the way that the data was laid out in the database. sometime it could be done, but an alternative approach would be three or four times easier.

So much effort had gone into constructing these detailed specs that there was a rigid expectation the team would be delivering this feature precisely how it was turned over to the team. It became painful to have to go back and explain that the request wasn't practical or feasible. Bringing an utterly impractical offer to a Development Team is devastating to team morale and is something you want to avoid at all costs.

The Agile approach is to bring the team a description about what is needed by the customer, and why this will be valuable. The team (sometimes with the help of architects) can then design a solution that will require less effort and provide greater value than one dictated to them by someone who doesn't understand how the code and the data has been constructed.

How about now - While working with teams at a company making tools for newspaper and magazine publishers, the Development Team I worked on came to dread production releases. Every six months, the team had one very demanding customer who would describe what was needed in very vague terms. The team had learned from past release complaints to bring prototypes. Everyone looked to this stakeholder to ensure there weren't any complaints during the release.

Unfortunately, the stakeholder would waffle back and forth, wanting minor changes here and there in a never-ending stream of small tweaks (which weren't improving the quality in any significant way). In the end, the feature would be released but took four times more effort than needed. For the developers, it was frustrating and a great source of uncertainty; you never knew if what you had produced would be good enough.

Exposed to the churn - Early in my career, I believed that I should share everything I learned with my Development Team. So, when I heard through the grapevine about possible new features or directions that the product might take I would share them. At companies (especially large companies), there can be a lot of politics and heated discussions around what features or new products will be worked on for upcoming releases.

One day, I learned a hard lesson while collecting status on a feature that a developer had started working on several weeks earlier.

Me: "How is the feature going?"

Dev: "Oh, I've done some work, but I'm spending most of my time helping Adrian with unit testing."

Me: "We're supposed to be "code complete" in four days, and this is a complex feature."

Dev: "Yeah, but it's not like this is going to be used."

Me: "What do you mean?"

Dev: "Well, they are already planning the new module, which would make this whole feature unnecessary, so why put any time into it?"

A Development Team needs to focus and collaborate to deliver quality features. While the drama and politics about what features will get built (mainly on the business and marketing side) is a necessary debate, it's an enormous distraction and demotivating for your Development Team.

Whom do you listen to - I worked as a QA Manager at a startup. We were working on the initial product to launch the company into business. Getting the features right was critical. Several Program Managers and a User Design Engineer roughed out the initial design. As we built layer upon layer of supporting functionality, questions started to pop up about handling things we hadn't foreseen in the initial design. We started bringing early looks of some of The Business's features to show our progress. We faced inconsistent feedback.

The Program Manager who had done much of the design liked what we had produced. The Program Manager's boss didn't like what we had created and wanted to change the workflow significantly. Sometimes the proposed changes made sense, but other times, they were very user-unfriendly. We didn't have a tiebreaker or a way to force them both into a room to reach an agreement. It is a challenge for Development Teams to know with whom to discuss situations with contradictory input.

Post-Release abuse - I've worked at places where Development Teams would hold their collective breath when a release came out and was first reviewed by The Business. Teams received criticism or hostility because something The Business wanted wasn't included or delivered the way they wanted. Sometimes there was a justifiable fault, but many times the request hadn't been communicated properly, had been communicated too late, or hadn't had enough detail to understand what was truly needed. When so much effort has gone into getting a software release out the door, it's awful, demoralizing and can lead to people leaving the company when the team gets criticized for something that they didn't know about.

The backseat driver - When complex systems get large enough, adjusting can require extensive and complex changes in the system's back end. Changes to the business logic, the layout of the database, or the middleware layer can require a lot of effort without much visible benefit. Program Managers or business leaders who outlined new features seldom had any patience or understanding for how difficult it could be to deliver new features.

For someone on the outside, the new feature might appear to need a single webpage and a few drop-down menus. When estimates come back saying that a new feature will require significant time and effort, it is frequently met with skepticism or outright pushback. This is incredibly demoralizing for Development Teams and an enormous distraction. Along with trying to break down and plan the work, the team now has to justify the effort it will take or attempt to shave off time (likely cutting into the quality of the work produced).

A surgeon would never be second guessed when he tells people how long a surgery is going to take, and we shouldn't be second guessing an engineering team when they give you an estimate.

PO support for Program Managers

Being a Program Manager in the traditional process was challenging because you were responsible for creating a schedule for what The Business wanted rather than what could realistically be completed by the Development Teams. Teams working on the features weren't typically notified before creating the schedule, so they had no input.

Program Managers are not part of the Development Team, so they weren't included in the most detailed work breakdowns and had a minimal view of the Development Teams. Program Managers' main focus was tracking work and putting pressure on the team to work faster.

Without an accurate inside view of the details of the project, Program Managers commonly got an overly optimistic view of the project progress from developers who were in the beginning to middle stages of the project. They then watched as the progress ground to a halt for the last ten percent of the project (which could take as much time as the first fifty percent took).

While working at a large multinational networking company, I was involved in a sizable cross-team project. A schedule had been created in which twelve weeks' worth of work would be completed by two groups, who would then hand off work to my team on week twelve. Once we got the code checked in, my team would be coding and testing for two, two-week Sprints before we could hand the final product off for final integration testing.

The project started with much fanfare and a beautifully complex gaunt chart. Weekly status meetings were held with crisp, precise, and detailed status emails sent out each week. As the project went on, the delivery of the first piece fell further and further behind. On the eleven[th] week, just one week before my team was to receive the code, the project was only finishing the milestones for week five. Despite the delays, the status email sent by the Program Manager was as crisp, precise, and detailed as ever, showing the project as green.

Puzzled, I asked how it could be green as there wasn't any way to turn it over to my team on schedule the following week. I was assured that everything was fine and this wasn't a problem I should be concerned with. I didn't think this was a perfect answer, so at the weekly status meeting, I brought up my concerns. Because the code was still seven weeks behind, my team didn't have anything to work with. My concerns were listened to carefully, but the status email continued to show the entire project as green.

At that point, I became vocal because the status had gone beyond being questionable and was now purely dishonest. But this isn't an uncommon problem. Even after my pushback, the Program Manager refused to send out a status that reflected where the project was.

I was told that she would do that, but only once a complete re-plan had been completed (which wasn't going to happen any time soon).

The project limped along for another couple of weeks, then shut down, and the Program Manager quit. Unfortunately, this wasn't and still isn't an uncommon problem. If the project had used an Agile approach, the delays would have been evident at the end of each Sprint when not enough functionality was ready for the Sprint Demo.

A common challenge was Program Managers equipped with high-level information about what the customers or The Business wanted but without the product background and knowledge to fully understand the trade-offs and details the Development Team needed to complete the work.

Key Take-Aways

- The Product Owner role was created out of a set of best practices

- Product Owners help Program Managers have an inside view of the progress that a Development Team is making (or not making)

- Being dedicated to the team and product area, Product Owners can do a better job of capturing the details and trade-offs needed to execute new features successfully

The Product Owner role for The Business

Business leaders don't have time to oversee work in Product Development; instead, they have a Product Owner there to represent their needs.

Provide The Business with an inside view into a Product Development team(s) - Not all organizations are set up with the Product Owner reporting to or even part of The Business, though this is how the role was initially set up. To do it well, you need to think of yourself as a representative for The Business. The Business needs someone on the inside to look out for their needs, to be their eyes and ears.

Ensure a strong return on investment (ROI) - While representing The Business, you are there to ensure a robust return on investment. You'll need to make decisions and judgments to prioritize work actively, so teams are capturing the core value of their work and not wasting time on feature refinements that don't provide significant value. (Think, good enough, not the massive time sink of perfection).

An on-the-ground customer advocate - As a Product Owner, you are always in touch and thinking about what your customer needs and how they will be using the product. You bring this view to all of your User Stories when you accept work, and you think about it when you look over new user interfaces or any other customer-facing content.

Management and a reliable view into feature risks - The Business needs someone they trust to mitigate and call out risks that might delay or impact new and upcoming features. As a PO, The Business wants to know you are trying to foresee future issues and minimize any impacts.

Product quality validation - The Business needs someone to ensure that before any User Stories and features are accepted, they will deliver the value they should to the end customer. The PO is needed to oversee the end quality of what is delivered.

To advocate for The Business, you need to be in close contact, be a confidant who is up to date about future initiatives and potential changes to priorities, and be trusted to give them an inside view of how product development is going.

To deliver for The Business, you need to keep an eye on the schedule and adjust keep it on track while keeping The Business up to date with any changes. You need to ensure a robust quality program within the team and that the features and products to be delivered are user-friendly and have quality built-in from the beginning. Providing quality sometimes means you will need to adjust User Stories and features because the original ask wasn't intuitive and wouldn't deliver the required customer experience.

As the inside person, you ensure The Business is getting what they need. When priorities change, you are the person who makes the proper adjustments, so the team switches directions to work on what is most important. When The Business needs to know if an important feature will be done on time, you can give them an honest assessment and facilitate adjustments.

Having a single, accessible point of contact for The Business is a revolutionary change, and probably the single most significant benefit Agile made for The Business. Now there is someone to coordinate with and ask questions of you, the Product Owner.

Key Take-Aways

- In the traditional model, The Business often had to deal with disappointing results and a lack of communication

- Late releases, low quality, and a lack of transparency for The Business often led to a lack of trust and confidence in the Product Development Process

- The Product Owner fills a key role by being an insider, providing visibility and transparency for The Business

- For The Business, the Product Owner is a single point of contact, defines the needed features, and accepts or rejects work completed by the Development Team

---

Product Owner support for The Business

A company's business and sales departments are vastly different from the product development world and operate under different rules and guidelines. The pressure to perform and sell is constant on the business side. With worldwide competition available at the click of a button, most businesses are marketing and sales-driven by necessity. Sales have direct access to the customers and are specialists who make money by engaging and selling, but only if they can close the next deal. With a short-term focus on commissions, quotas, and making sales, business units don't have time to focus on product development.

Here are some of the issues that The Business had with Product Development.

Timeliness - Important customers developed new features or changes they wanted with the original product. The Business needed to get these changes completed quickly. But with the traditional process, there wasn't anyone to gather up the details and watch over the work through the development process (especially for bug fixes or other items that needed to be fixed quickly). These shortcomings commonly lead to needed features not showing up quickly or only partially delivering what was required. This was compounded by the long cycle times of a traditional release process. To understand the gravity, let's step back and look at this problem in action.

Say that your Product Development Process is working on a six-month release process (short for a traditional approach).

The first six weeks of the release were spent building out detailed requirements documentation, an ultra-detailed plan was produced (down to how many coffees breaks per day), and the teams are now two months into a six-month process.

The government has come up with a hot new privacy change that your top customer now needs, and they want to know when you can release it. The easiest possibility is to wait until the next release and add it as the highest priority for the next six-month release. Of course, adding the change into the next release means you would need to finish the current four months and add it to the next six-month cycle. The change won't get in front of a customer for ten months (in the best-case scenario, which is unlikely). Your customer won't be happy and doesn't want to wait for Option One so let's look at Option Two.

Option Two takes the carefully constructed plans and schedules for the current release and scrambles them to add in work needed for the new feature. This will be a bit messy because this new work will completely invalidate the existing schedule, and the carefully constructed technical requirements will be challenged as you add this change into the middle of an intricate multi-layered technical design. The schedule was already tight, and it's unclear how much of an impact this work will have on the commitments already made for the existing release. But getting it in front of a customer in four months instead of ten is often the choice made.

The quality of the releases to the end customer was also an enormous challenge. Quality challenges showed up in a multitude of forms. Common among them:

User hostile features - Features could deliver The Business's end functionality but have confusing text, an overly complicated layout, or unclear instructions. Customers were guaranteed to be unhappy with the results.

Features with incomplete testing - Quality Assurance is at the end of the Software Development Lifecycle, so when development goes longer than it should, and the end customer expects to have a functionality by a specific date, it's the quality that suffers. Quality issues commonly showed up as transactions that error, unstable transactions that would fail intermittently, and lots of rushed fixes for problems that went unnoticed before they impacted customers in production.

Priorities out of synch - Interaction with large important customers happens because The Business needs to understand the requirements of the highest revenue accounts. During the six to nine-month extended software release, The Business's priorities and the needs of sizable important end-customers continue to evolve and change.

The Business's priorities changed months ago, and the original version is no longer valid. There can be so much momentum around the initial priorities for a Product Release that priority changes don't always get appropriately updated at the team level. Accepting that effort is no longer needed when so much effort has already gone into delivering a feature, particularly when there isn't enough time to complete the new priority effectively, can be a struggle for teams.

Low return on investment - The Business must make hard choices allocating capital to maximize their return on the investment. When properly advertised and with a top-notch sales team, even a mid-level product can make significant money, so taking money out of the marketing and sales loop to invest in future capabilities is a considerable risk. When a product's quality and customer value drop enough, the top-level salespeople can't make the money they expect and will leave (likely to a competitor who is doing better). To keep a company on a firm footing, The Business has to get a good return on the product development budget they invest. Getting a low return on capital robs momentum and allows competitors to pull ahead.

You talked to whom - Finding the right person to talk to when changes in direction were needed or when The Business wanted an update on a new feature was challenging. Work on the Development Team wasn't divided into teams, and there wasn't a guide for what developer was working on what feature. Most of the development work had become highly specialized. Because the work has been divided among three-six different development groups with a similar (but less complex) problem on the QA site, a feature might be worked on by a User Interface Designer, a Back-End Developer, a Web Developer, an Architect, a Software Engineer in Test, and a Quality Assurance Engineer.

Want to know how the new feature looks and how it's working? Good luck finding out!

The Product Owner role for the Development Team

Having a PO has simplified life for Development Teams by providing them with a single, direct communication line, allowing them to build and test new features without chasing down information or people. Here is how successful Product Owners play within the Development Team.

A single point of contact to understand what is needed - As an insider to The Business, the PO helps the Development Team understand what The Business needs. This is accomplished through User Stories, which lead to questions, conversation, and clarification. Having one place to ask questions and get feedback is critical and allows the Development Team to increase its speed and quality dramatically. When new requests or features are needed, having a knowledgeable Product Owner who represents what change The Business needs is critical and eliminates wasted effort.

Shields the team from distractions and churn - Good Product Owners shield the Development Team by filtering out the drama and the politics as The Business debates what features will get built and which customers to focus on. The PO decides what features and priorities to work on in the current and future releases. The Development Team is free to focus on execution and not be affected by the churn and stress inherent in the business processes.

Ensures the relevancy and completeness of new requests of the team - The Product Owner acts as a Gate Keeper for new requests and ensures the Development Team doesn't see any until they have been prioritized high enough. The high-priority requests will likely be worked on soon and have enough information in the User Stories to make them actionable.

Protects solution ownership - The Product Owner protects the Development Team's ability to own and develop the best solution possible. To do this, the PO must listen carefully to what The Business is asking for, capture and clarify what is being asked, and call out and remove the details asking for a particular solution. This allows the team to design/select a solution that will deliver what is needed with the best quality and lowest level of effort.

Accepts work from the Development Team - By accepting work from the Development Team, the Product Owner allows the team to deliver quality content rather than worry about the work being rejected by The Business or expending extra effort trying to deliver more than what is needed.

An advocate - As an insider, the Product Owner has a unique ability to provide insight and confidence in the work and efforts that the Development Team is putting out. As someone familiar with the difficulties and the work required to deliver content, the PO can help stand up for the team and inspire confidence that the company is getting a good return on investment.

Key Take-Aways

- The Product Owner acts as a single point of contact to detail and answer questions about what the customer needs

- The Product Owner shields the team from organizational churn and politics so that they can focus on delivering and testing features

- The Product Owner ensures that the team only sees requests that have enough detail, and are of a high enough priority that they should be worked on

- The Product Owner protects the development team's ability to design or co-design the solution with help from the architects

- The Product Owner reviews, accepts, and marks as done work completed by the team

- The Product Owner acts as an advocate for the development team, helping outsiders and the business understand the difficulties and effort that the development team undertakes

PART 6 - ADVANCED STRATEGIES

A deep dive into Sprint Demonstrations

As the Product Owner, you should be the Master of Ceremonies for your team(s). A good Sprint Demo is a chance to showcase your mastery and vision for your product area, the team's accomplishments, and what to expect in the coming Sprint. Set the expectation with your team that a Sprint Demo will be delivered at each Sprint. They need to be thinking about what they want to show off and who will be doing it (rotating responsibility is a good idea).

In your role as PO, it's crucial to establish a regular, steady drumbeat of success. Even if your stakeholders can't attend your Sprint Demos, it is still critically important for the Sprint Demonstration to happen every Sprint. Remember that when the developers on your team look good, so do you. Be sure to give your team plenty of time and spotlight to let their work shine.

Because you want to create a regular, repeatable cadence of success, it works best (and is less work) if you create an excellent

Sprint Demo template and an email that you send out and re-use every time. Senior Managers, Vice Presidents, Engineering Directors, and other executives need demonstrable success and progress. If you send out a regular flow of Sprint Demo emails, these can be available and used to establish a positive reputation for your team and your Product Owner skills.

Here is the information you want to build into your Demo Template.

Page 1

Team Name

Date

Page 2

Sprint Goal

Why did you have this Sprint goal?

The work/features this supports

Page 3

Risks you were watching

Any new or adjusted program risks

Page 4

Intro to the fantastic and talented developers on your team and the work they completed this Sprint

Page 5

Great things accomplished by team members and thank you's

Page 6

Sprint goal for the upcoming Sprint

Why this Sprint goal is important

What feature or initiative this is supporting

Any dependencies

Key Take-Aways

- As the Product Owner, you should be the person out front giving the Sprint Demonstration

- Make sure to give a demonstration every Sprint

- Send out a recap email after every demonstration and keep the same format every time

- Use the template

Hire a public relations firm

First, let's ignore the question about how to hire PR help and ask instead if money wasn't a problem, "Wouldn't it be great to have your very own Public Relations Firm. They could help highlight your accomplishments, the big wins, the behind-the-scenes things that were great, but that people typically don't hear about?"

You see it all the time in the corporate world. Minimal talent and work ethic latch onto wins (often those they didn't contribute much to) and wave them around to get applause and kudos. Your team is doing significant work and making progress; why wouldn't you want to get proper credit for accomplishing it?

It's a classic recurring theme that people, especially developers, don't get the recognition they deserve. All too often in their careers, they watch Program Managers and other people stand up to get the credit, applause, and bonuses in company meetings when the Development Teams made it happen. When you've put in that much work and see someone else get credit, it stings no matter who you are.

Not being recognized is consistently top of the list for employees unhappy with their workplace; unfortunately, it's widespread. The great news is, it's a straightforward fix and has a massive payoff.

When your team gets the recognition it has been missing, it pays off on the following:

- Personal level – the person who put in the great work gets recognition for their efforts giving them a feeling of accomplishment and gratitude towards you for recognizing what they've done

- Team level – Your team will see you're looking for, recognize, and will voice when they do great things, and this will encourage them in the future to want to step up

- Project level – The leaders and other teams on your project will see your team's positive work and will want to level up their work to match

- Organizational level – Budget and headcount are assessed and handed out regularly. Executives and managers need to ensure resources are being used efficiently and to good effect, and when vital work needs to get done, they want successful teams to take it on. When your team has a consistent, repeatable drumbeat of success, you're going to be at the top of the list for headcount, budget, and exciting work

- Bonus – Just as your team needs PR, your boss and their boss need PR too. It would be best if you had a steady, regular stream of successes so that the wins for your team turn into wins for your boss and your boss's boss

- Bonus Bonus – As the designated servant leader for your team, when your team looks good, is working well, kicking butt, and taking names, then you will look great. You look much better than if you try talking up your accomplishments

Okay, so maybe I've convinced you that you need a PR firm. But getting back to reality, how will you fit that in? Even a lousy PR firm is likely beyond your budget.

Well, no one knows the wins and inside story as you do, and there is always the time, money, and energy for documenting successes. It's essential. The good news is, you and the Scrum Master can do it, and it won't take much time at all. The hardest part is getting into the habit of paying attention and calling out the good things that have happened. I'll show you how to set up a system to make this easier.

PR isn't a one-time thing. Make sure the praise is justified and done in a regular repeating cadence. It needs to be a consistent message to stick, not something piled on heavily once in a while. Invest in your team by taking the time to set this up.

First, create a demo email template so you can easily send out a recap of all of the great things presented at each Sprint Demonstration. Give the email template a catchy title, attach it to a reminder, and send it to yourself or whomever the Product Owner is for a given team.

Next, create a PowerPoint template for your Sprint Demo. Make this template look good because you will be using it for a long time and if it looks good, you and the team will look good.

Here are the things you'll want to include:

- The team's focus and the business objectives for the Sprint
- Team accomplishments
- Focus and objectives for the next Sprint
- Risks and dependencies for the next Sprint
- Team members
- Who is making the Sprint and what it's showing

"What if we're not doing demos? What if we do demos, and no one shows up? What if we do demos, but no one uses PowerPoint?" Well, if any of these are the case, congratulations, you're about to be a rock star by starting!

Key Take-Aways

- Getting your team credit for its accomplishments is critical for its long-term success

- Successful, productive teams get more resources, people, and exciting projects

- Your boss and the boss of your boss also need to show accomplishments

- Developers get grumpy when other people get credit for their work

- Grumpy developers aren't as motivated or productive

When to include your friends

If you are one of the multiple teams working in similar and connected areas, it helps to do cross-team planning. Numerous frameworks focus on Agile at Scale, including Less and Safe, to name only a few. While these frameworks have differences, they are similar in that they hold significant, multi-day, cross-team planning events, commonly called Big Room Planning events or BRP.

At BRP events, teams cross coordinate to understand if anyone depends on the work they will be doing and to plan out the next six to eight Sprints. The goal is to ensure everyone understands and accepts any cross-team dependencies and set a realistic, achievable plan to deliver the most value possible without taking on work that can't get completed.

To maximize value and the ability to produce a quality plan, your team must come prepared. Your team will be asked to create a Sprint-by-Sprint plan describing the User Stories they will commit to completing at these events.

This will include breaking epics/features into User Stories, understanding the stories enough to give them realistic estimates, and crafting them into a Sprint-by-Sprint plan. Unless the work your team is producing is ultra-simple, it's not going to be possible to complete this within the short time you'll have at BRP.

If you don't come prepared, you won't have time to connect with the other teams to complete the User Stories required to unblock your team. If you have broken down these epics/features, you'll be able to go to the other teams before they have constructed their Sprint-by-Sprint plans and get your work included.

A Warning: The Safe Agile framework clings to the outdated idea that teams shouldn't come to BRP events prepared with epics/features broken down or pointed, and it's true that if you do all of the work beforehand, your team won't be as engaged, but unless your work is ultra-simple there just isn't enough time to do everything you need to do in these events.

The reality is that these events are messy. There will always be last-minute changes and teams coming to you with stories they need to get into your backlog. Your initial plans will get modified by other team requests and priority changes, so don't get too invested in the first versions. If you come prepared, you're going to be able to move things around, concentrate on the new features or stories, and work through the changes. If you don't come prepared, there won't be time to participate in cross-team coordination. And that is the real value of Big Room Planning.

Key Take-Aways

- Big Room Planning events are an excellent way for teams and Product Owners to collaborate and align their efforts

- During Big Room Planning events, there will always be some changes that teams will need to make based on the collaboration required by other teams and gaps found during the event

- Come prepared to the Big Room Planning event by having features broken down into stories and the stories roughed out and pointed

- Coming prepared will allow your team the space to adjust and point to new features and work without taking on massive amounts of stress and churn

Enlist allies

Regardless of how your company is organized, Development Managers should work alongside the Product Owner to help the team succeed. A good Dev Manager can have a significant positive impact so it's best to work closely with them because they can be valuable allies.

Dev Managers will have a deep background in the team's current and historical strengths and weaknesses, and know past performance.

It's vital that you ask for and listen carefully to their opinion. By listening, you're showing that you value their opinion, that you genuinely want to work together to help the team improve, and ultimately their opinion matters (so you'll want to take notes). Of course, taking interest and notes won't be enough if you don't act.

Make an ally and a partner; reserve some time and ask them, "What areas do you see as key to the team's success and why?" The aim is to create a positive partnership to improve and support the team, so be prepared to share your assessment.

Along with a general assessment, most managers have critical areas they tend to focus on based on memorable career failures or successes. They know at a core level that these areas must be working well for anything to be successful. Even if everything else is going right, these managers may key in on their hot button topics.

Having the Dev Manager as an ally means you have someone else to help you change behavior and make improvements. An ally is a second pair of eyes and ears, another mind to bounce ideas off of or to help you think them through.

Getting an ally won't do much good if you don't keep them, so you'll want to follow through on that partnership. If things are going well, this may be something you only need to do occasionally, but if the team is struggling, you'll want to be collaborating to make progress.

Here are some ways to foster a strong working relationship:

Share information - When you hear about a program, project, or other impactful changes, share what you've heard with your allies. They may not have heard the information you have or may have additional details that you don't, so keeping each other in the loop helps you prepare for changes that will affect the team.

Share insights - As you discover new things or notice weak spots for the team, share those with your allies. They may have additional information or unique approaches to addressing things you may not have considered.

Support/acknowledge their hot-button issues - Mention or give visibility into ways you and the team are improving or focusing on your allies' hot-button issues. You can do this with a quick mention or reference on how a fix or improvement will also help out in the hot-button area.

Ask for help - Everyone wants to be needed and valued. Asking for help is significant for your ally. By doing so, you demonstrate your belief in their ability to provide support and advice, helping them feel good about what they are doing.

Key Take-Aways

- The development manager is responsible and interested in the personal growth and development of their people. They are an invaluable partner in the software development process

- Having allies will be crucial to your success as a Product Owner, and everyone else's success in their roles

- Providing value and keeping your commitments are critical to maintaining allies

- Providing information, insights, and support for your allies and their causes encourages them to do the same for yours

- Utilize your allies and ask for help when you need it

Support a winning approach

If your team is going to be successful, having a healthy culture is a must. A common language to help make the team a team, so don't underestimate the importance.

Here are the powerful messages you'll want to convey regularly and model boldly.

Message - No one is perfect, and no one has all the answers.

Why is this important? To get the team fully engaged, you need them to know that you don't have all of the answers, and you don't even have all of the answers when it comes to the product (even though it's what you're being paid to do). Even if you feel that you have all of the answers (because you're just that good), you may not have factored in all of the information for a given situation, or there may be an underlying problem you're not aware of. When there is more space for discussion, there is more space for questions to come up. If the team understands that you are open and expect to learn and collaborate with them the relationship will have a healthier footing.

Message - We are smarter when everyone participates.

Why is this important? Teams typically only have a few senior members. If the less senior team members hold back and let the senior people take the lead with questions, code, and design ideas, you will only be utilizing twenty-thirty percent of the team's brainpower. Senior members are a real asset because of their experience and product knowledge. Junior people are also a vital asset because they typically have strong skills in newer technologies and tools. Without understanding, they can ask pointed questions about "why do we do it that way?" which is excellent for uncovering trade-offs made years ago that might no longer make sense. It's going to be critical that everyone participates fully.

Message - This is a team sport.

Why is this important? When people on the team start thinking of the team's success above their own, you get smarter decisions and support that doesn't happen until that type of thinking kicks in. It also helps them appreciate the work that other people do on the team, which helps morale.

For example, I had a team where a senior person was great at getting around build issues and getting people unstuck so they could get back to coding. During a month, multiple changes in the build process brought people to a standstill five or six times. If Siva (name changed) had been worried just about his productivity or his pride, he could have continued on his way. Fortunately for us, Siva was a team player, and with his knack for troubleshooting, he would sit down with team members and together they would fix in twenty minutes what would have taken them two or three hours to do.

Message - We help and support each other.

Why is this important? A culture of mutual support makes it acceptable, even expected, for people to ask for help and for help to be offered, thereby saving valuable time when a team member runs into problems. It also encourages teaming, leading to better results as each person can provide valuable experience and compensate for others' weaknesses.

Message - This is a great team; I'm happy to be heard; I'm honored to be working with you.

Why is this important? When your team's chemistry and morale are high, the output, the effort people are willing to spend, and the enjoyment and dedication to the team all go up significantly. Every team has strengths and weaknesses, and everyone wants to feel pride in their work. When you take pride and appreciate being on the team, so will they.

Message - If anyone speaks up when they think something is not right, they will not be ridiculed if wrong, or ostracized if they are right. (Psychological Safety).

Why is this important? In order to harness the intelligence and full capabilities of a team everyone has to feel safe to be able to speak up and add their contributions. Without this Psychological Safety the team reverts from being a team and becomes an extension of whom ever is leading ruining the whole value of having a multi-talented and capable team.

Message - It is safe to make mistakes.

Why is this important? Innovation and working smart doesn't happen without taking some risks. If your team doesn't feel safe risking a mistake they won't innovate or improve your product.

Key Take-Aways

- Team and organizational culture are the foundation upon which everything is built

- Without a healthy culture, you don't have a team; you have a group of people, some pulling in different directions

- Success requires the entire team to be involved

- Teamwork requires psychological safety

- Critical decision-making can't be reserved for a few individuals if you want the team to succeed

- As the Product Owner, you are a team leader and have significant influence

---

Being in the organizational loop

To serve your team, you and your Scrum Master need to be privy to the latest information about where the product and company are headed. Some data may be preliminary, but it's still worth knowing.

"Difficult to see, always in motion the future is." Yoda

The company/product/feature may end up taking a different direction, but having a heads up is critical as it allows you and the Scrum Master to start thinking about these changes. The changes may or may not fit into current efforts/systems/architecture. Warning may allow you to proactively bring up concerns or hidden costs other people may not have considered. Being in the loop about where your product might be headed can help you make smarter choices today, saving you time and headaches in the future.

Recently, I heard that an automation tool that had been the standard for years in our organization had become overly complicated with too many drawbacks.

A replacement tool was being considered. Our team had come to the same conclusion a quarter earlier but hadn't been able to get enough people to back an alternative at the time. Armed with new information, I gave my senior test engineer a heads up. Because our team had already reached this conclusion, we had already made significant progress towards a better alternative.

Our engineers were able to find and engage with the newly appointed group tasked with a replacement, show them the benefits and progress we had already made, and get our solution adopted across the program (instead of seeing the future go in a different direction).

Key Take-Aways

- Being connected to organizational knowledge gives you and the Scrum Master advance warning about potential future changes

- Understanding potential changes allow you and the Scrum Master to prepare and influence direction, saving your team time and effort

---

Building team capital

Team capital is goodwill, trust, and camaraderie that turns a collection of people into an integrated unit. Great teams, like marriages, don't happen overnight; no matter how good the chemistry is, it takes time to establish the trust, communication, and roles that will make it successful. Because team capital isn't easily measured on a graph or with a number, it's often ignored by much of the business world. Still, like a good marriage, team capital produces impressive results.

When measuring success, productivity, and happiness, poor team capital can be just as devastating as a divorce (at least for the team's existence). There isn't a number we can use to assess a healthy relationship, "We've been at a six point two for three months, and she won't do anything about it! Time to get a divorce lawyer." But that doesn't make it any less real or impactful. Let's take team capital seriously and acknowledge its importance.

Disagreements are logical differences of opinion and something that should be discussed and weighed in a healthy team setting. In the normal day to day operation of a team there will be many disagreements and it's important that the team can work through them in a healthy manner. When there is plenty of team capital in the bank to pay for the withdrawal that a disagreement brings, the team can move beyond these and continue communicating, collaborating, and rebuilding trust.

Without enough team capital disagreements can change into arguments which step beyond logical differences and become emotional. Arguments aren't helpful and consume team capital because they can't be solved, often end with perceived winners and losers, and seldom have useful outcomes.

As team capital gets low, small disagreements will become arguments and will start impacting team members' willingness to work together. Technology is a collaborative team sport. In extreme cases, conflicts or deep-rooted distrust can take hold and rip teams apart, and when that happens, almost nothing productive gets done, and quality will also suffer. I know that you don't want to get anywhere close to that type of scenario, so let's invest some capital.

Hosting a team-building event once a quarter can be good for morale but some people really hate the idea that the company is trying to force team building. As a one-time event, the slight bump in team capital will have been long since spent by the time the next quarter rolls around. The best way to build capital with little effort is to make a lot of small deposits. Frequent deposits tend to compound on each other, and they can build up surprisingly quickly. It works better to not call these events "team building events", but to make them into rewards for the team that they are enjoying together.

Here are two foolproof ways I have found to do this.

Build capital method 1- Food and cooking together

Until recently, we as humans only ate with our family or tribe, so meals are a built-in way to bond and build trust. Cooking together might not be practiced regularly, but it's an activity that only close friends or family tend to do together. And naturally, after you cook together, you eat together.

See if your workspace can accommodate a lunch table to allow people to sit together for meals. If it can't, scheduling a regular reoccurring team lunch is a great alternative. People don't tend to turn down free food.

I like to make waffles or crepes for breakfast when the teams have done an excellent job for a release. We're eating together and building the team capital, and it shows my and the company's appreciation for the work and effort the team has put out.

Potlucks are always a fantastic way to build morale and team capital. Our workforce is diverse, and many of your team members may have a different heritage. I like potlucks because people from different cultures often have a rich, exceptional culinary heritage to bring to the table. When a tasty dish is shared with

co-workers, the entire team can celebrate and appreciate this difference.

Build capital method 2 - Daily questions

Another great way to build team capital is to add simple daily questions that each team member can answer. These are easy to add to the daily Standup, add about five minutes, and over time the answers to these questions provide you with a deeper understanding of your co-workers.

I like to use Table Topics, which has an inexpensive set of icebreaker cards, such as "What was your favorite meal as a child?" or "What did you want to be when you grew up?" If you use these, just be sure to review the questions beforehand because some, such as

"Who was your first kiss?" could make some people uncomfortable or inappropriate and should be removed from the deck before you use it. Some seemingly innocent questions might trigger something in someone's past, so this method works best if questions are optional and anyone can pass on a question without the need for an explanation.

There are lots of different ways to build team capital but however you do it the end result is worth it.

Well-functioning teams are significantly smarter, and more productive than the sum of the individuals in them. Here are a couple of websites that have some great ideas.

1. https://www.themuse.com/advice/team-building-activities-games-for-work-office

2. https://www.workamajig.com/blog/team-building-activities

Key Take-Aways

- Teams are greater than the sum of their parts

- Team capital is what holds a team together

- Invest early when creating a new team to create team capital (and get it going)

- Regular small investments are the best way to increase team capital

- Events that are seen as rewards or celebrations will be better received than "team building events"

Make an ally of your Scrum Master - sidestepping conflict

The Scrum Master is one of the three pillars that make up a functioning Agile Team, so the role is critical for your success as a Product Owner. No matter how good a Product Owner is, if the team doesn't support a good Scrum Master and isn't following Agile well, it will be hard to succeed. If the Scrum Master and the Product Owner interact so much that if they are not getting along it's going to be a major distraction to the team so it's much more productive to have a good partnership.

The Scrum Master will usually have a lot of insight into how well the team is functioning, and ways that it might work better. There will be plenty of times when the Scrum Master won't make a Standup, or a Product Owner has to skip a User Story Grooming meeting to meet with The Business, so it's essential to have a strong partnership so each role can help cover for the other.

Here are detailed talking points that will serve as a good starting point in creating this partnership. There is a certain amount of flattery because you'll want the Scrum Master to feel confident in their role and comfortable that you accept them and their abilities, but you need to mean what you say.

Introduction to the topic - Thanks for coming. Since we're going to be working so closely together, I thought it would be helpful to meet to understand each other's background; we can talk about how best we can support each other and how best to partner while helping the team.

I am interested in you - So, why did you become a scrum master? What do you like about it? What makes it difficult? What is your most successful moment as a scrum master?

I'm thrilled that you're here - I'm good at understanding the product, but for the team to be successful, we need someone who understands Agile. I'm delighted that the team has you as our Scrum Master.

Let's partner - I know that there will be times when I am stuck in meetings with The Business and won't be able to attend User Story Grooming sessions or other ceremonies. I know that there will be times you won't be able to make Standup or might have a conflict. Hence, I thought it would be helpful if we partner so that if that happens, we can cover for each other, and the team won't have a gap. If there is ever a ceremony that you're going to miss, just let me know, and let me know how I can best represent your views, priorities, or the changes that you need.

My goal and role - As I see it, my primary goal is to make sure that you, as the Scrum Master and the team, have the details to be successful. No matter what else happens, I haven't done my job if I haven't done that. To ensure success, I'll provide the details, the User Stories, and prioritization, and you can provide the expertise in Agile and the Agile ceremonies. Does that sound like a plan?

Get buy-in - Do you have any thoughts on how we can partner? Does this sound like a reliable approach?

Key Take-Aways

- The Scrum Master Role is critical to the success of the Agile Team

- To make things work, you want to make an ally and work closely with your Scrum Master

- Scrum Masters and Product Owners need to back each other up and fill in for each other as needed

- You want to ensure that the Scrum Master doesn't see you as someone trying to compete with them but rather as someone helping them

---

PART 7 - WHAT'S NEXT

Congratulations by reading this book you've taken a bold step forward to improving your Product Owner skills, your career, and the success of your company. Let's take a look at the content that we've gone over.

We talked about every Agile ceremony that the Product Owner will be attending, how to deliver, and what to avoid.

We detailed the value of well written User Stories, how to refresh out of date stories, and then gave you the information and a template to help you deliver them.

We walked through the origin of the Product Owner role, and how the Product Owner interacts and solves the needs of the Business, the Development Team, and the Executives.

We detailed how to get your team the visibility and recognition it deserves by delivering top notch Sprint Demonstrations and providing Public Relations for your team.

We talked about how and why to avoid short term thinking, having QA outside of the team, poor quality, and untracked work.

We talked about how to build collaborative and successful relationships with the key people that partner with the Product Owner, and within the Development Team.

With practical definitions we talked about how the Definition of Done and the Definition of Ready are key documents that are needed in the day to day operation of an Agile Development Team.

The last area we touched on was to talk about some of the roles and strategies that successful Product Owners employ like why to shield their Development Teams from drama and churn.

The next step for you is simple.

Decide if you want help, or if you want to go it alone, but roll up your sleeves and get started.

Don't over-think this, and don't try to improve too many things at one time. Make decisions, make improvements, and keep going.

The transformation won't happen overnight, it will take time, but I've given you everything that you need to be a Rock Star Product Owner. You just need to get started.

How to get more help

If you do decide you want our help transforming your career, or your organization, just reach out and let's sit down and chat use the link to schedule a time with us.

calendly.com/advancedagileexecution

We're good at what we do just like you're good at what you do we've had the time to become experts in the product owner space and if you have something to offer the world or your organization then we want to be the ones to amplify it for you.

To your success,
Matthew Kramer
and the Advance Agile Execution team

ABOUT THE TEAM

Advanced Agile Execution is a collaboration between Matthew Kramer and Tanja Diamond. Our mission is to help you go from being an average Agile user- to an Advanced Agile BOSS!

Matt started his tech career in Seattle, WA in 1998 working his way up from QA Engineer, QA Lead, SDET, QA Management, Program Manager, Scrum Master, and then an Agile Coach. Matt has worked in small startups, Fin-Tech, and Fortune 100 tech companies.

In 2018 Matt completed the construction of a two-thousand square foot barn which started with the initial layout of the foundation, framing, plumbing, and finished with the wiring (he did get a little help with the wiring).

When he isn't building barns, furniture, or Agile Programs, Matt volunteers to help young adults land their first tech jobs, and grows tree starts to distribute them in the community.

Tanja Diamond is a Maverik leader and has been a Business and Life Strategist for over 38 years. When she isn't out leading innovation in her clients' lives, she is saving lost and injured animals.

www.ingramcontent.com/pod-product-compliance
Lightning Source LLC
Chambersburg PA
CBHW071554200326
41519CB00021BB/6737